T0185184

Solutions Manual
to Accompany

Hydraulic Engineering

Second Edition

John A. Roberson
Professor Emeritus
Washington State University

John J. Cassidy
Consulting Hydraulic and
Hydrologic Engineer

M. Hanif Chaudhry
Professor and Chairman
Civil and Environmental Engineering
University of South Carolina

John Wiley & Sons, Inc.
New York • Chichester • Weinheim
Brisbane • Singapore • Toronto

ISBN 0-471-24496-1

10 9 8 7 6 5 4 3 2 1

CHAPTER TWO

2-1. Flood Control Vol. = 210000 AF = 210000 AF (43560 SF/A) = 9147.6 million cu ft
= 9147600000 AF (0.028317 cubic meters/cubic ft)=259.033 million cu m

Consv. Vol. = 323100 AF = 323100 (43560) = 14074.24 million cu ft
= 14074240000 (0.028317) = 398.540 million cu m

2-2. 1310 cu ft/sec = 13 10 cu ft/sec (0.028317 cu m/ cu ft) = 37.1 cu m /sec
= 1310 cu ft /sec (3600 sec/hr)(24 hr/day)/43560 s ft/acre =2598.3 AF/day

---

2-3. a. Vol = 13 in(35 sq mi) = 455 sq-mi-inches
= 455(640 A/ sq mi)/(12 in/ft) = 24,266.7 AF
= 455(5280 ft/mi)/(12 in/ ft) = 1,057,056,000 cu ft
= 1057056000 (0.028317 cu m/ cu ft) = 29,932,654.8 cu m

b. Runoff = 24,266.7 (0.7) = 16,986.7 AF
Q = 16,986.7/72 = 235.9 AF/ hour
= 235.9 (43,560 sq ft/A)/3600 sec/hr = 2,854.7 cu ft/ sec

2-4. Annual Flow Vol. = 40(3600)(24)(365)/43560 = 28,958.7 AF/yr

86000/28958.7 = 2.97 yrs

2-5. Loss = (40 - 21)(1000)/12 = 1583.3 AF/ yr

Years to fill = 86000/(28958.7 - 1583.3) = 3.14 yrs

2-6. Obtain precip in inches from Fig. 2-2.

Precip (in)

	1959	1960		1959	1960
J	1.3	0.55	J	0.50	0.12
F	2.3	1.06	A	1.70	1.00
M	0.34	2.43	S	3.33	1.15
A	1.82	1.14	0	2.00	0.80
M	1.00	0.70	N	0.23	1.30
J	1.45	0.67	D	1.00	0.82

Sum of 24 months = 28.71 in
Avg. = 28.71/24 = 1.2 in/ mo   Max. Dev. is Sep 1959 = 3.33-1.2 = + 1.3 in

---

2-7. The probability of a flood overtopping levee in any one year is

$P = 1/75 = 0.0133$

The probability of the levee being overtopped twice in 60 yrs is

$P = (n!/((k!)(n-k)!))(p^k)(1-p)^{(n-k)}$ (Eq. 2-4)

where n = 60 yrs

k = 2

Solving Eq. (1) for P yields <u>P = 0.144</u>

2-8. a) $P = 1/T =$ <u>0.050</u>

b) Let "success" mean no floods in 3 yrs then P(success) + P(failure) = 1 where "failure" means at least 1 twenty year flood in the 3 yrs.

Thus P(success) = 1 - P(failure)

where $P(\text{failure}) = 1 - (1-p)^n$ (Eq. 2-6)

or $P(\text{success}) = 1 - (1 - (1-p)^n)$

$= 1 - (1 - (1-.05)^3)$

$= 1 - .143$

= <u>0.857</u>

c) P(failure) = 0.143 (from above)

d) from (Eq. 2-4)

$P(1 \text{ in } 3) = (3!/(1!(3-1)!))(0.05)^1(1-.05)^2$

= <u>0.135</u>

e) 10% risk = P(failure) = P(1 or more floods)

or $0.1 = 1 - (1-p)^n$ (from Eq. (Eq. 2-6)

where n = 5

Solving for p yields p = 0.02085

or T = 1/p = <u>48 years</u>

2-9. 25% risk means $P = 0.25$
$n = 30$ years, $P = 1-(1-p)^n$ (Eq. 2-6) $0.25 = 1-(1-p)^{30}$
solving for p yields $p = 0.00954$
But $p = 1/T$, Thus, $T = 1/p = 1/0.00954 =$ 105 years

2-10. $Q = M + KS$, $S = 110$ cfs, $M = 750$cfs, $p = 1/T = 1/100 = 0.10$
From Table 2-3, $K = 2.326$, Thus, $Q = 750 + 2.326(110) =$ 1006 cfs

---

2-11. Assume a hometown of St. Louis, Missouri. Fig. 2-8 shows that average precipitation is approximately 37 inches. For Phoenix, Arizona Fig. 2-8 indicates an average annual Precip. of approximately 10 inches.

2-12. From the data we find the mean of $Q = 3{,}154$ cfs $= M$. Also the standard deviation is $S = 1896$ cfs.

a) For the normal probability distribution : $Q = M + KS$

T (yrs)	Exceedance Probability	k (Table 2-3)	Q (cfs)
10	0.10	1.282	5,585
50	0.02	2.082	7,101
100	0.01	2.326	7,564

For the Pearson III distribution we must use the logarithms of the discharges given. Using the common logarithms of the Qs in the table gives the following values.

$X = \text{Log } Q$, $X = 119.772$, $X = 412.549$, $X = 1429.505$, $N = 35$

Thus, using Eq. 2-7 $M = 119.772/35 = 3.422$

Eq. 2-10 -- $S = [((412.549 - (119.772)^2)/35/34]^{1/2} = 2.81$

Eq. 2-11 -- $g = [((35)^2(1429.505) - 3(35)(119.772)(412.549) + 2(119.772)^3]/[35\text{x}34\text{x}33)(0.281)^3] = -0.871$

Using $X = M + KS$ the following table is constructed:

T (yrs)	Exceedance Probability	k (Table 2-4)	X	Q (cfs)
10	0.10	1.152	3.746	5,572
50	0.02	1.566	3.862	7,278
100	0.01	1.681	3.894	7,834

b) Eq. 2-6 -- $P = 1 - (1-p)^n$ where $n = 3$ years, $P = 1/T = 1/20 = 0.05$
$P = 1 - (1 - 0.05)^3 =$ 0.143

2-13. From adding the values in the table, total 12 month precip = 36.6 in

a. Total Vol. = 36.6(110) = 4026 sq-mi-in = 4026(640)/12 = 214,720 AF

b. Calculate monthly flow volumes. Example, for Oct: = 37(3600)(24)31/43560 = 2275 AF

Month	Flow Vol. (AF)	Precip. (in)	Net Precip. (in)	Runoff %
O	2275	12907	-10560	17.6
N	2975	25227	12907	11.8
D	3751	38133	34613	9.8
J	676	39893	37547	1.7
F	1000	34613	31680	2.9
M	4919	29333	23467	16.8
A	20826	18187	8800	114.5
M	28592	8213	-2933	348.1
J	11960	2347	-10560	509.6
J	11375	1173	-15253	969.7
A	9039	1173	-15253	770.6
S	4760	3520	-8213	135.2
Totals =	102148	214719	86242	

Total Flow Vol. for Year = 102148 AF

c. Example for Oct: Precip. Vol. = 110(2.2)640/12 = 12,907 AF
Net Precip. Vol. = (2.2-4.0)110(640)/12 = -10,560 AF
All months are completed in this way in above table.
Total Net Precip. for Year = 86,242 AF

d. Example for Oct: % Runoff = 12275/12907 = 0.176 = 17.6%

% Runoff for Year = 102148/214719 = 47.6 %

e. Evapotranspiration continues and uses up soil moisture.

f. Flow out of the basin includes ground-water contribution which came from infiltration during earlier rains.

2-14. P(2-hr) = 20 in. Vol. = 20(10) = 200 sq-mi-in = 200(640)/12 = 10,667 AF

2-15. a. P(avg) = (3.2+2.8+4.1+1.6+2.4)/5 = 2.82 in.

b. Thiessen Polygon Method: See the constructed figure from which the following areas have been planimetered.

Station	Area (sq mi)	P (in)	PA (sq-mi-in)
A	5.83	3.2	18.66
B	12.09	2.8	33.85
C	6.50	4.1	26.65
D	13.99	1.6	22.38
E	5.59	2.4	13.42
Totals	44.00		114.96

P(avg) = 114.96/44 = 2.61 in.

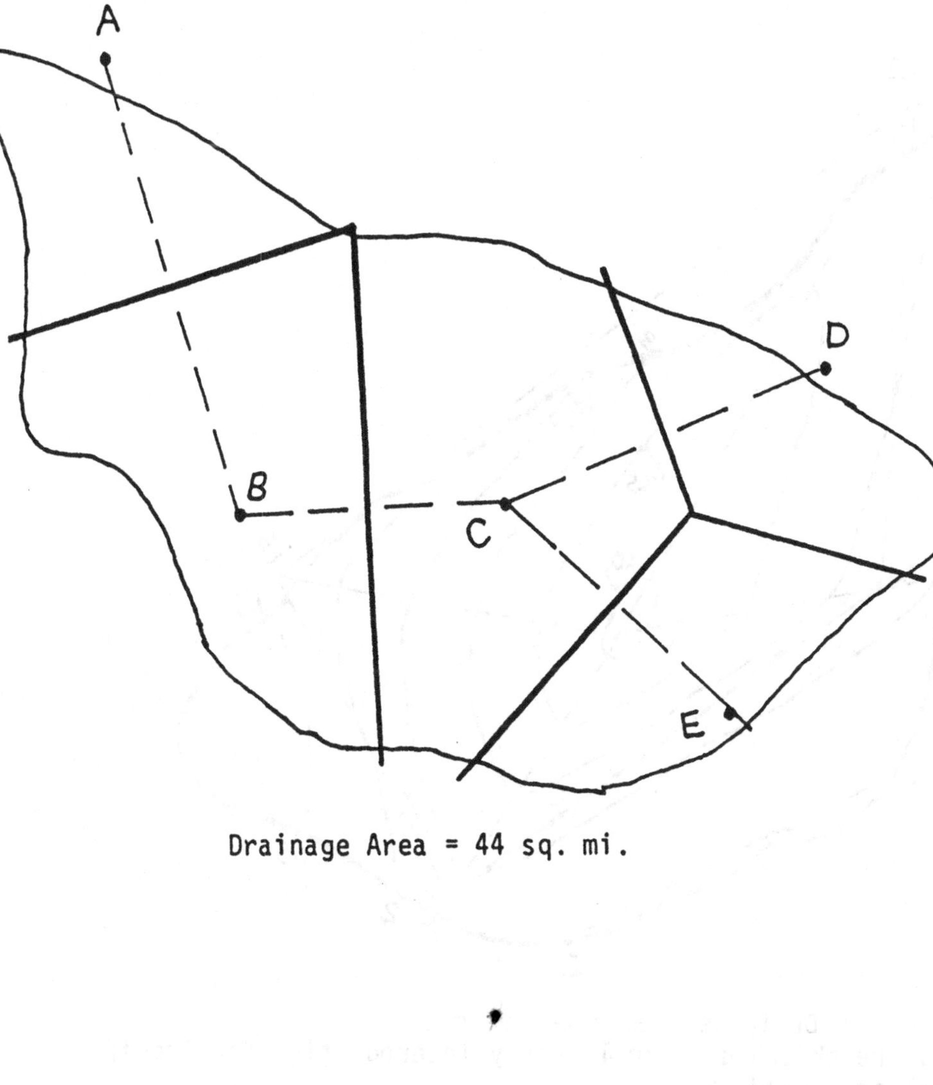

Drainage Area = 44 sq. mi.

2-16. See the constructed isohyetal maps. The areas and average precip. values have been taken from the maps. The areas were planimetered. There will be variation between individual maps.

a.

Isohyet	Area Between (sq mi)	Pav (in)	PA (sq-mi-in)
4.1-4.0	0.26	4.05	1.05
4.0-3.5	6.60	3.75	24.75
3.5-3.0	9.44	3.25	30.68
3.0-2.5	12.33	2.75	33.91
2.5-2.0	10.52	2.25	23.67
2.0-1.5	2.79	1.75	4.88
1.5-1.0	2.06	1.25	2.58
Totals	44.00		121.52

Pavg = 121.52/44.00 = 2.76 in.

b.

Isohyet	Area Between (sq mi)	Pav (in)	PA (sq-mi-in)
4.1-3.5	12.75	3.8	48.45
3.5-3.0	11.90	3.25	38.68
3.0-2.5	10.00	2.75	27.50
2.5-2.0	5.56	2.25	12.51
2.0-1.5	2.53	1.75	4.43
1.5-1.0	1.26	1.25	1.58
	44.00		133.15

Pavg = 133.15/44.00 = 3.03 in.

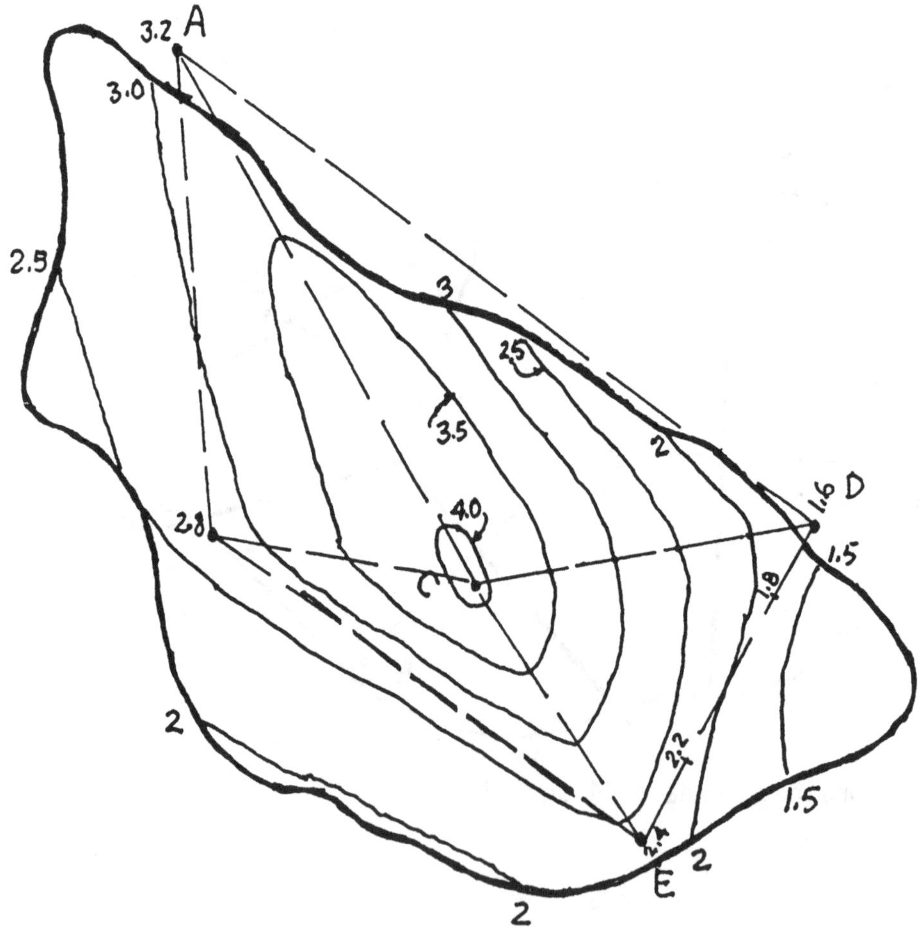

Drainage Area = 44 sq. mi.
Isohets are sketched after linearly interpolating for Precip. depth along dashed lines.

Prob. 2-16 Cont'd.

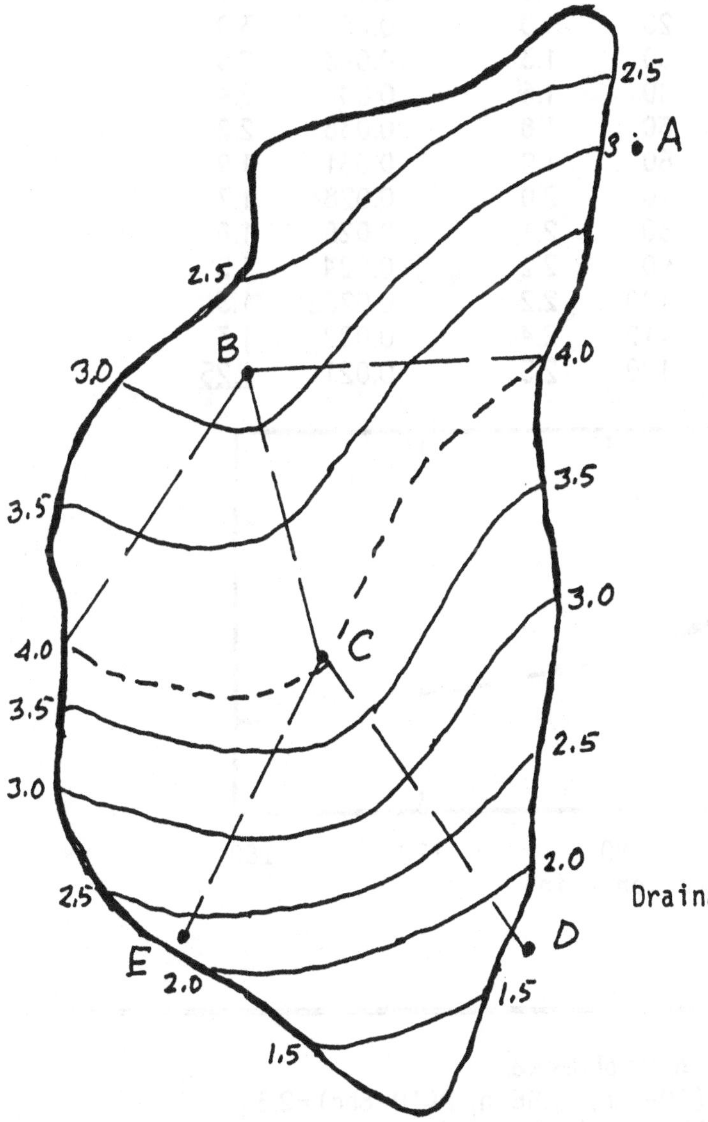

Drainage Area = 44 sq. mi.

Isohyetal Map with 4-in. Isohyet given. Other isohyets are sketched after linearly interpolating along dashed lines shown and keeping more or less parrallel to 4.0-in. isohyet

2-20.

Time	Accum. Depth (in)	Incr. Depth (in)	Duration (min)	Max. Depth (in)	Intensity (in/min)	Intensity (in/hr)
12:15	0					
12:25	0.1	0.1	10	0.6	0.06	3.6
35	0.2	0.1	20	1.0	0.05	3.0
45	0.4	0.2	30	1.3	0.043	2.6
55	0.7	0.3	40	1.6	0.04	2.4
13:05	1.0	0.3	50	1.8	0.036	2.2
15	1.4	0.4	60	1.9	0.031	1.9
25	2.0	0.6	70	2.0	0.028	1.7
35	2.1	0.1	80	2.1	0.026	1.6
45	2.2	0.1	90	2.2	0.024	1.5
55	2.2	0.0	100	2.2	0.0232	1.3
14:05	2.4	0.2	110	2.4	0.022	1.3
15	2.5	0.1	120	2.5	0.021	1.25

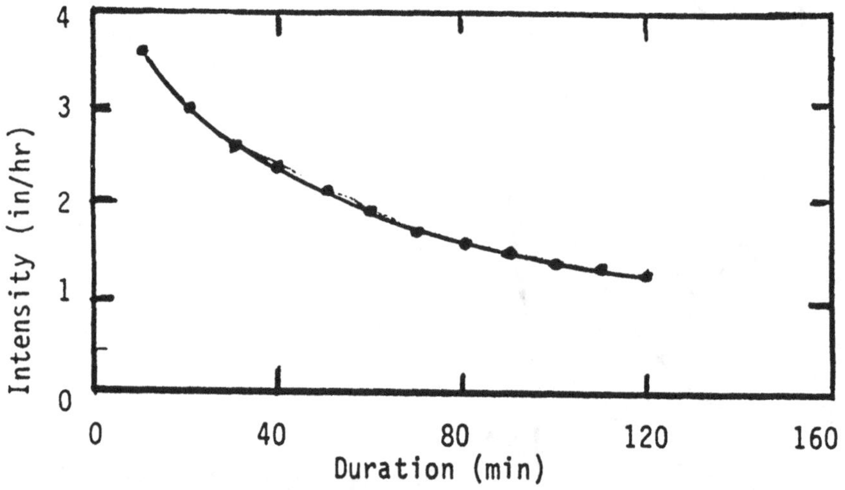

2-21. Assume a location at the S.E. corner of Nevada.

From Figs. 2-17,18,19: P(10-1hr) =1.56 in, P(10-6hr) =2.3 in

P(10-24hr) = 3.1 in

From Table 2-5: P(2-24hr) = 3.1/1.60 = 1.94 in

P(1-24hr) = 0.78(1.94) = 1.51 in

P(5-24hr) = 1.40(1.94) = 2.72 in

P(25-24hr) = 1.82(1.94) = 3.53 in

P(50-24hr) = 2.00(1.94) = 3.88 in

P(100-24hr) = 2.22(1.94) = 4.31 in

These values are plotted on the accompanying graph. A straight line is drawn through the 10-yr values and straight lines are drawn parrallel to the 10-yr line through the 24-hr values for 1,2,25,50, and 100-yr values. These are point precipitation values. The values in the following table were read from the curves for the indicated durations.

Prob. 2-21 Cont'd.

Values Accumulated depth for each duration have been read from the previous table and plotted on the following graph of Precipitation depth versus duration.

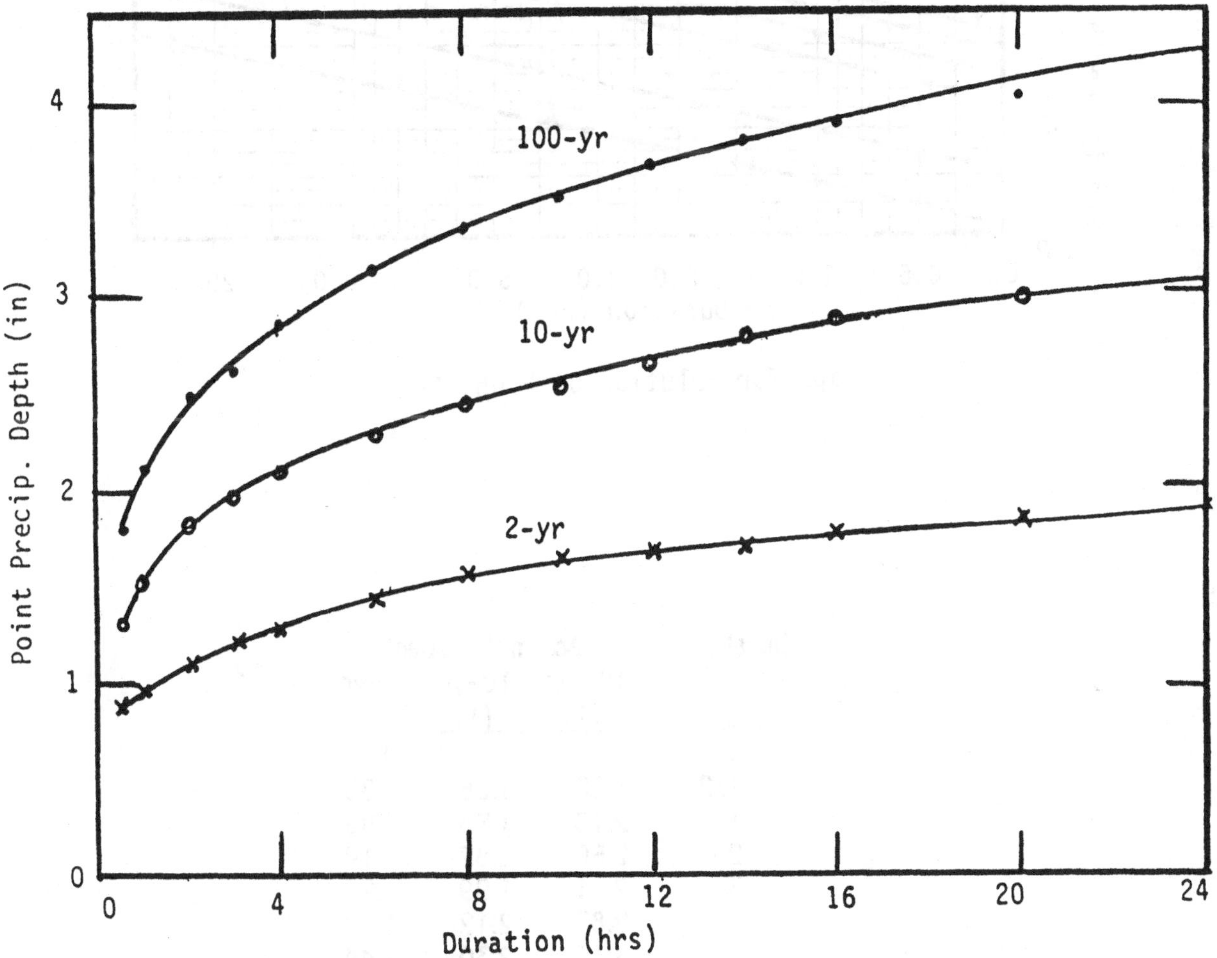

Prob. 2-21 Cont'd.

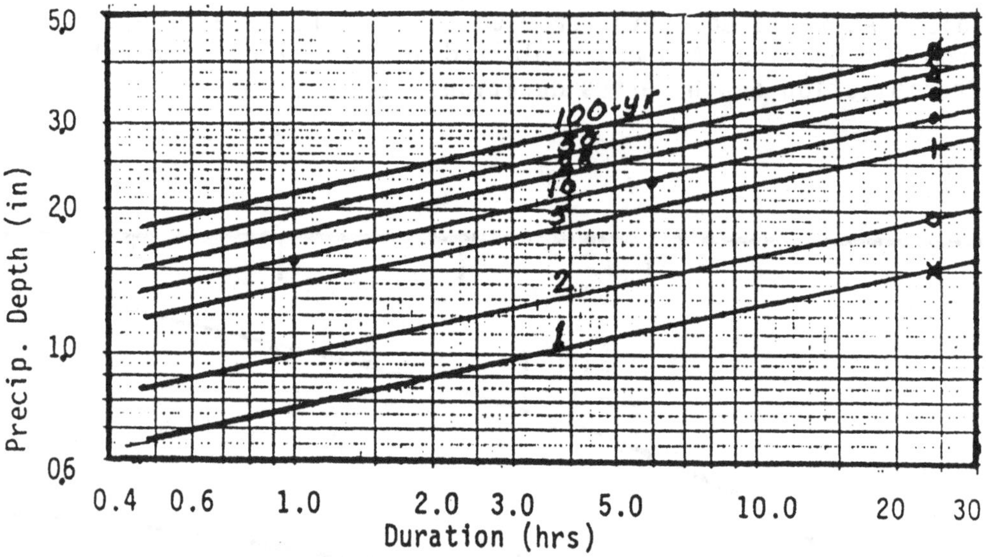

Graph for Solution of Prob. 2-21

Duration	Accumulated Depth		
	100-yr	10-yr	2-yr
(hrs)	(in)	(in)	(in)
0	0	0	0
1/2	1.85	1.36	0.85
1	2.13	1.56	0.98
2	2.50	1.82	1.14
3	2.71	1.99	1.25
4	2.87	2.12	1.32
6	3.15	2.30	1.44
8	3.36	2.48	1.54
10	3.51	2.60	1.62
12	3.67	2.70	1.68
14	3.80	2.79	1.74
16	3.92	2.87	1.80
20	4.08	3.00	1.88
24	4.31	3.10	1.94

2-22. The following values have bee read from the graph drawn for the solution for Problem 2-21.

Duration (hrs)	Accumulated Depth (in)	Duration (hrs)	Accumulated Depth (in)
1	1.56	7	2.40
2	1.82	8	2.48
3	1.98	9	2.55
4	2.10	10	2.61
5	2.21	11	2.66
6	2.31	12	2.70

2-23. The values of incremental precip. were computed from the table in Prob. 2-22 and have been rearranged with the peak value 1 hour after the midpoint and the remaining values arranged in descending order around the peak as suggested in the text.

Time	Point Incr. (in)	Area Avg. (in)	Px/P12 Type 1	P Type 1 Accum. (in)	P Type 1 Incr. (in)	Px/P12 Type 2	P Type 2 Accum. (in)	P Type 2 Incr. (in)
0	0	0	0	0	0	0	0	0
1	0.04	0.04	0.035	0.09	0.09	0.022	0.06	0.06
2	0.08	0.08	0.076	0.20	0.11	0.048	0.13	0.07
3	0.09	0.09	0.125	0.33	0.13	0.080	0.21	0.08
4	0.10	0.10	0.194	0.51	0.18	0.120	0.31	0.10
5	0.12	0.11	0.515	1.35	0.84	0.181	0.47	0.16
6	0.26	0.25	0.682	1.79	0.44	0.663	1.74	1.27
7	1.56	1.51	0.767	2.01	0.22	0.820	2.15	0.41
8	0.16	0.15	0.830	2.17	0.16	0.880	2.30	0.15
9	0.11	0.11	0.878	2.30	0.13	0.915	2.40	0.10
10	0.07	0.07	0.926	2.43	0.13	0.950	2.49	0.09
11	0.06	0.06	0.963	2.52	0.09	0.975	2.55	0.06
12	0.05	0.05	1.000	2.62	0.10	1.000	2.62	0.07

Sum = 2.70 x.97=2.62 in.

The area avg. values for 22 sq mi were obtained by multiplying the point values by 0.97, the area adjustment factor obtained from Fig. 2-21 for 22 sq mi. The Px/P12 factors were obtained from Table 2-7. However, because we are interested in a 12-hour storm instead of a 24-hour storm the time values have been halved, preserving the distribution of the storm but cutting the duration in half. The Incr. Precip. depths for the three storm distributions have been plotted in the following graph.

Prob. 2-23 Cont'd.

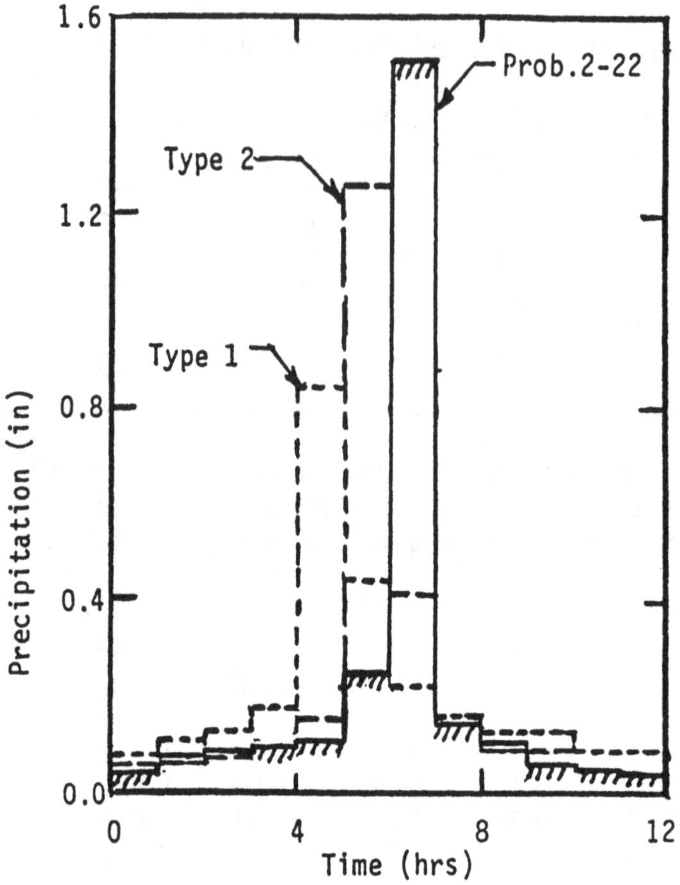

2-24. Example for Chicago, Illinois
From Fig. 2-17,18, & 19,
P(10-1hr) = 1.97 in, P(10-6hr) = 3.00 in, P(10-24hr) = 4.00 in
From Table 2-5: P(100-24hr)=2.22[P(10-24hr)]/1.60 = 2.22(4.0)/1.6=5.55 in

These values of P are plotted on the accompanying figure. The accumulated Precip. values of column 2 of the following table were read from the graph. An area correction factor of 0.92 was read from Fig. 2-21 for a drainage area of 200 square miles and used to adjust the point values of column 4 to the 200 sq. mi. values of column 5 for Prob. 2-25.

Prob. 2-24 Cont'd.

Duration (hrs)	Accum. P (in)	Incr. P (in)	Point Storm Incr. P (in)	Area Adj. Storm Incr. P (in)
0	0	0	0	0
2	3.22	3.22	0.10	0.09
4	3.80	0.58	0.15	0.14
6	4.20	0.40	0.20	0.18
8	4.45	0.25	0.20	0.19
10	4.65	0.20	0.25	0.23
12	4.85	0.20	0.58	0.53
14	5.05	0.20	3.22	2.96
16	5.20	0.15	0.40	0.37
18	5.30	0.10	0.20	0.18
20	5.40	0.10	0.10	0.10
22	5.50	0.10	0.10	0.09
24	5.55	0.05	0.05	0.05
Totals =		5.55 x 0.92=5.11		5.11

The values in column 4 are the selected sequence for the point storm sequence in Prob 2-24. The area adjusted values are the storm sequence adjusted for a 200 sq.-mi. area for Prob 2-25.

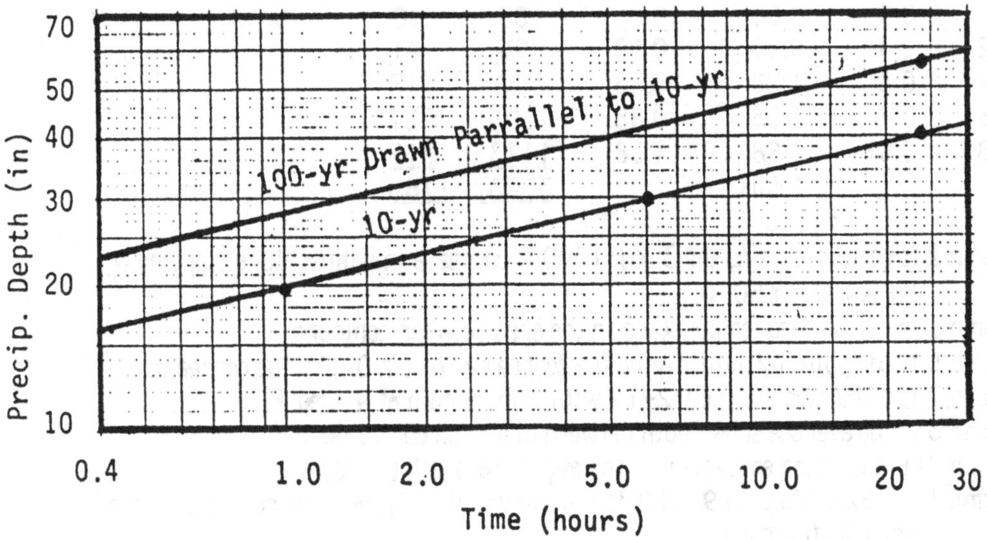

2-25. The table developed for Prob.2-24 was used to develop the storm for Prob. 2-25. See the notes on Prob. 2-24 where the point Precip. storm sequence was multiplied by an area adjustment factor of 0.92 to form the storm sequence for this problem.

2-26. From Fig. 2-25 Avg. Annual Lake evaporation at Lake Mead (east of Las Vegas Nevada in Arizona) is approximately 82 inches.

Annual Water Loss (evaporation) = 82(162700)/12 = 1,111,783 AF

2-27. Column 1 of the following table was taken from the table for Prob. 2-27.

Elev. (ft)	Temp. (deg F)	Avg. Temp. (deg F)	Incr. Area (sq mi)	Degree Days	Temp. Contr. To Melt (in)	Melt Due to Rain (in)	Avg. Snow Depth (in)
10600	34.2						
10000	36	35.1	5.5	3.1	0.31	0.13	61
9000	39	37.5	5.1	5.5	0.55	0.13	57
8000	42	40.5	9.5	8.5	0.85	0.13	54
7000	45	43.5	4.5	11.5	1.15	0.13	52.5
6000	48	46.5	15.5	14.5	1.45	0.13	46
5000	51	49.5	11.1	17.5	1.75	0.13	36

Snow Water Equiv. (in)	5% of Water Equiv. (in)	Total Melt (in)	Runoff (in)	Total Melt (AF)	Runoff (AF)
18.3	0.9	0.44	0	129	0
17.1	0.9	0.68	0	185	0
16.2	0.8	0.98	0.18	496	91
15.7	0.8	1.28	0.48	307	115
13.8	0.7	1.58	0.88	1306	727
10.8	0.5	1.88	1.38	1113	817
		Totals =		3536	1750

Column 2 is computed using the 3 deg/1000 ft decrease in temperature starting with 51 degrees at 5000.

Column 3 is the average temperature in the area between contours.

Column 6 is computed by multiplying the degree day factor 0.1 by the degree days.

Column 7 is calculated using Eq. 2-17 with 2 in. of rain at 41 deg. F.

Column 8 is the average snow depth between elevation contours.

Column 9 is the water equivalent of the avg. snow depth (30%).

Column 10 is 5% of Column 9. This is the amount of melt the snow can hold before any runoff occurs.

Column 11 is the total of the temperature and rain-produced melt.

Column 12 is the difference between the total melt and column 10, the portion of melt which will be held in the snow.

Column 13 is the volume of runoff calculated as total melt x drainage area x 640/12.

2-28. First determine total volume of runoff for the week.

$$\forall = \sum Q \Delta t$$

$$= (39.2 + 302.4 + 952.0 + 739.2 + 470.4 + 336.0 + 72.8)\ ft^3/s$$

$$\times\ 24\ hrs \times 3600\ sec/hr$$

$$= 251.6 \times 10^6\ ft^3$$

depth of runoff = V/A

where $A = 21.0\ mi^2 = 21.0 \times 5{,}280^2\ ft^2 = 585.45 \times 10^6\ ft^2$

depth of runoff = $251.6 \times 10^6 / 585.45 \times 10^6 = 0.430$ ft

= 5.157 in.

No. of degree days = (33 + 38 + 49 + 43 + 39 + 38 + 33) - 7 x 32

= 49

Avg. degree day factor = Depth of runoff/degree days

= 5.157 in./49 = 0.11 in./degree day

2-29. Total depth of rainfall = 3.80 in. (from amounts given in rainfall table).

Equivalent depth of runoff = V/A

= 19 A.F./120 acre

= 0.158 ft

= 1.90 in.

Thus the equivalent depth of infiltration = 3.80 - 1.90 = 1.90 in. By careful inspection of the rainfall table one can deduce that the value for the ϕ index will be greater than 0.20 in. and less than 0.40 in. The solution for ϕ is given as

$$\phi = 1.90\ in. - (2 \times 0.20\ in.)\ /4\ hr = 0.375\ in./hr$$

2-30. No runoff will occur for the hourly periods when the ϕ index is greater than or equal to the rainfall. Thus, no runoff occurs for the 1st, 2nd, 3rd, 7th and 8th hourly periods of rainfall.

Thus $R_0 = (0.6 + 0.40 + 0.40) - 3 \times 0.20 = 0.80$ in.

$V = (0.80\ in./12\ in./ft) \times 100\ acre = 6.67$ A.F.

2-31. Total precipitation depth is 8.1 inches (sum of values given in table for Prob. 2-31).

Vol. Runoff = 100(12)/200 = 6 in. Total loss = 8.1 -6.0 = 2.1 in.
Phi index = 2.1 in / 7 hrs = 0.3 in/hr

2-32. Use Eq. 2-19. $f = fc + (fo - fc)\exp(-k\,t)$
From Table 2-8 fc = 0.32 (avg. rate for clay loam).
Integrate Eq. 2-19 over the period from 0 to7 hours as follows:

$$\text{Vol.} = \int [0.32 + (1.5-0.32)\exp(-K\,t)]\,dt = 0.32(7) - (1.18/K)[\exp(-7k) - 1]$$

$$= 6.0 = 2.24 - (1.18/K)[\exp(-7K) - 1]$$

We solve this equation for K by trial and error as follows:

Assumed K	R.H.S.	
1.0	1.18	assumed K too large
0.1	5.94	assumed K still too large
0.095	6.03	assumed K O.K.

Thus, fo = 1.5 in, fc = 0.32 in , K = 0.095

2-33. From Fig. 2-22, P.M.P.(24-hr) = 27 inches for 200 sq mi
Fig. 2-23 shows that for Zone 7 for 36 sq. mi.
P.M.P.(48-hr) = 1.28 (27) = 34.6 in

2-34. From Fig. 2-22, PMP(6hr) = 0.86 PMP(24hr-200 sq mi) = 0.86(27) =23.2 in
PMP(12hr)=0.97 PMP(24hr-200 sq mi) =0.97(27) =26.2 in
PMP(48hr)=1.28PMP(24hr-200 sq mi) = 1.28(27) = 34.6 in
These values are plotted on the following graph.

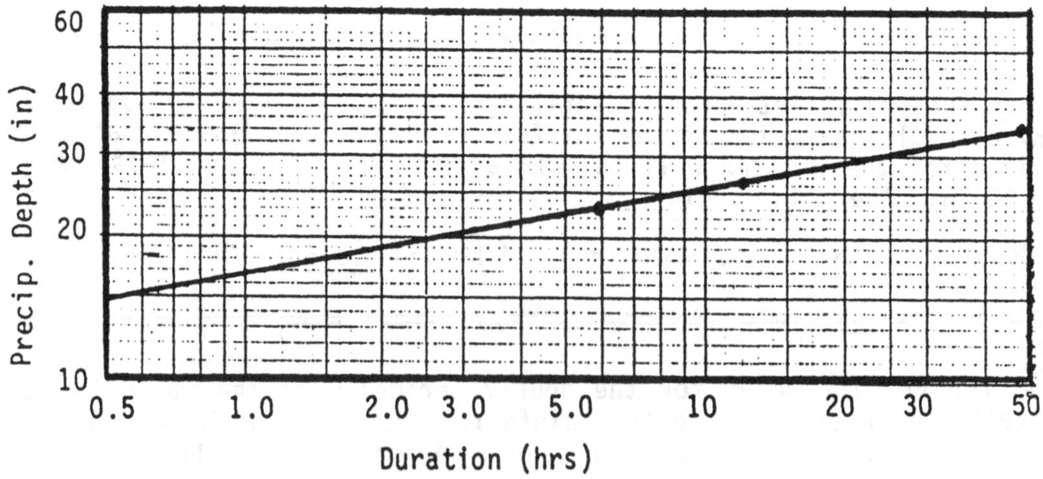

Prob. 2-34 Cont'd.

The following values of accumulated precip. have been read from the preceding graph.

Time (hrs)	Accum. Precip. (in)	Incr. Precip. (in)	Storm Sequence (in)	Time (hrs)	Accum. Precip. (in)	Incr. Precip. (in)	Storm Sequence (in)
0	0	0	0	26	30.6	0.5	2.5
2	19	19	0.3	28	31.1	0.5	19.0
4	21.5	2.5	0.3	30	31.6	0.5	1.7
6	23.2	1.7	0.4	32	32.0	0.4	0.9
8	24.5	1.3	0.4	34	32.4	0.4	0.8
10	25.4	0.9	0.5	36	32.8	0.4	0.6
12	26.2	0.8	0.5	38	33.1	0.3	0.5
14	27.0	0.8	0.5	40	33.4	0.3	0.4
16	27.8	0.8	0.5	42	33.7	0.3	0.3
18	28.5	0.7	0.7	44	34.0	0.3	0.3
20	29.1	0.6	0.8	46	34.3	0.3	0.3
22	29.6	0.5	0.8	46	34.3	0.3	0.3
24	30.1	0.5	1.3	48	34.6	0.3	0.3

The storm sequence has bee arranged with the peak slightly after the midpoint and the remaining incremental values arranged in descending order on each side of the peak. There is uncertainty in reading the graphs involved.

2-35. From Fig. 2-23 the PMP for a 200 sq mi area is approximately 31.8 in. Then for Zone 8 in which this area is located the percent factor to be applied to the 31.8 in. for a 400 sq. mi. area and a 12-hr storm is 0.80 (from Fig. 2-22). Thus, the desired PMP = 0.80(31.8) = <u>25.4 in.</u>

2-36. From Fig. 2-19, P(10yr-24hr) = 3.10 in
From Table 2-5, P(50yr-24hr) = 2.0(3.1)/1.6 = 3.88 in point rainfall
From Fig. 2-21, P(50yr-24hr) = 3.88(0.92) = 3.57 in, avg. over 200 sq mi.
From Table 2-8, Soil is Group C
From Table 2-11, CN = 86 (pasture, poor condition, Condition II)
From Table 2-12, CN = 97.2 for Condition III
From Fig. 2-28, for P = 3.57 in, rainfall excess = 3.1 in,
Volume of Runoff = 3.1(200)640/12 = <u>33067 AF</u>

2-37. P(50yr-24hr) = 3.57 as determined in Prob. 2-36.

a. From Table 2-11, CN = 79 (pasture, fair, Condition II)
From Table 2-12, CN = 93.4 for Condition III
From Fig. 2-28, Rainfall Excess = 2.9 in
Runoff Volume = 2.9(200)640/12 = <u>30933 AF</u>

b. From Table 2-11, CN = 74 (pasture, good, Condition II)
From Table 2-12, CN = 90.2 for Condition III
From Fig. 2-28, Rainfall Excess = 2.6 in
Runoff Volume = 2.6(200)640/12 = <u>27733 AF</u>

2-38. A. Storm Construction

1. Construct the storm using Figs. 2-17, 2-18 and 2-19. The data from these figures is given below:

Duration (hrs)	R_f (in inches)
1.0	0.80
6.0	1.50
24.0	1.90

When the above data are plotted for purposes of interpolation we have:

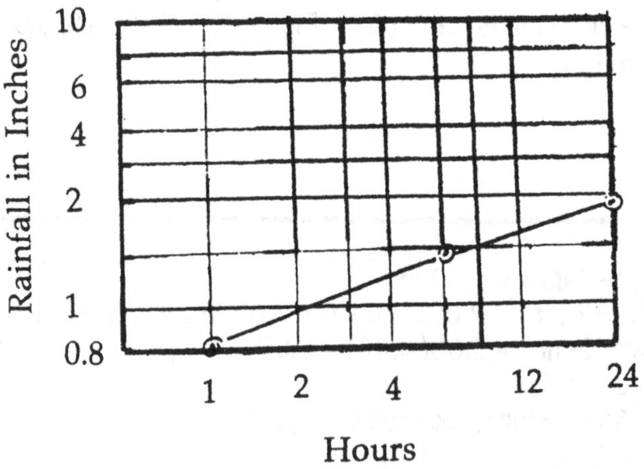

2. From the above graph tabulate the accumulated precipitation from the time of beginning of the storm to 24 hours later. Note: a considerable amount of approximation is involved in this process. These accumulated values are given in column 2 of the table (below). Next, determine increments of rainfall for every 2 hours and rearrange them to yield a representative rainfall sequence. Then use an area factor of 0.97 (from Fig. 2-21) to develop the storm rainfall. These three sets of data (rainfall increments, re-arranged rainfall increments, and rainfall increments adjusted for area) are given in columns 3, 4 and 5 (below).

Time (hrs)	Accum. R_f (inches)	Increments (inches)	R_f sequence (inches)	R_f (adjusted for area) (inches)
0	0.		0.	0.
2	0.97	0.97	0.05	0.05
4	1.20	0.23	0.05	0.05
6	1.50	0.30	0.23	0.22
8	1.55	0.05	0.97	0.94
10	1.60	0.05	0.30	0.29
12	1.65	0.05	0.09	0.08
14	1.68	0.03	0.05	0.05
16	1.71	0.03	0.04	0.04
18	1.80	0.09	0.03	0.03
20	1.84	0.04	0.03	0.03
22	1.87	0.03	0.03	0.03
24	1.90	0.03	0.03	0.03

B. Determination of CN values.

1. The soil was described as "clay like"; therefore, assume that we have soil group C (from Table 2-8).

Prob. 2-38 Cont'd.

2. Evaluate CN.

Land use	%	CN value	Adjusted CN Type III moisture--Table 2-12
Wheat (small grain)	70	84	96
Pasture	25	86	97
Roads (dirt)	5	87	97

CN (avg.) = 0.70 x 96 + 0.25 x 97 + 0.05 x 97 = 96.3 ≈ 96

C. Determination of Rainfall Excess (surface runoff)

The determination is done in tabular form below. The rainfall excess is obtained by entering Fig. 2-28 with rainfall.

Time (hours)	R_f (inches)	Accum. Precip. (inches)	Rainfall Excess (inches)
0	0.0	0.0	0.0
2	0.05	0.05	0.0
4	0.05	0.10	0.0
6	0.22	0.32	0.02
8	0.94	1.26	0.85
10	0.29	1.55	1.20
12	0.08	1.63	1.24
14	0.05	1.68	1.25
16	0.04	1.72	1.28
18	0.03	1.75	1.30
20	0.03	1.78	1.32
22	0.03	1.81	1.36
24	0.03	1.84	1.40

Approximation is involved in reading these vaues from Fig. 2-28

2-39. In the following table, the surface inflow volume, direct precipitation, and evaporation loss for each month has been calculated as follows: (example for Mar.)

Surface Inflow = 190 cfs days(2 AF/cfs-day) = 380 AF
Direct Precip. Volume = 3.1 (620)/12 = 160 AF
Evap. Loss = 2.9(.7)62 0/12 = -105 AF
Change In Vol. = 380 + 160 - 105 = + 435 AF

Month	Surface Inflow (AF)	Direct Precip. (AF)	Evap. Loss (AF)	Change In Vol. (AF)
Mar	380	160	-105	+435
Apr	168	108	-166	+110
May	164	186	-170	+180
Jun	14	145	-184	-25
Jul	42	98	-188	-48
Aug	18	67	-184	-99
Sep	58	31	-148	-59
Oct	34	46	-108	-28
	Total Change in Vol. =			+466 AF

2-40. Use Rational Equation Q= CIA   I (intensity) = 3.6/(3 + .25) = 1.1 in/hr
From Table 2-9, C = 0.82 (Avg. value for concrete pavement)

Q = 0.82 (1.1)20 = 18.0 Acre-inch/hr
= 18.0 cfs

2-41. Assume base Flow = 100 cfs. The volume of surface is given as
Vol. = (1/2)(12 hrs x 3600 sec/hr x 800 cu ft/sec)
=17.28 million cu. ft.
Depth of runoff = Vol./ A = 17,280,000/(5280)(5280)(10)
= 0.06198 ft = 0.744 ft
Peak Discharge for unit graph = 800 / 0.744 = 1075 cfs

2-42. From Prob. 2-36, Total Precip. = 3.57 inches.
This total must be arranged in a reasonable storm sequence. Use Type 2 distribution from Table 2-7. Determine t_c for the drainage area:

Eq. 2-22, $t_c = [(3.35 \times 10^{-6}) L^3 / h]^{0.385} = 593$ minutes = 9.9 hours

Use 10 Hours

Calculate required increment in duration: $t_r \leq 10/5 = 2$ hours
Thus, we use 2-hr increments of rainfall and a 2-hr unit graph. Form the following table using the Type 2 factors from Table 2-7. The rainfall excess in the last column is obtained from Fig. 2-7 using CN = 97.2 as determined in Prob. 2-36.

Pbob. 2-42 Cont'd.

Time (hr)	Type 2 Factor Px/24	Accum. Precip. Depth (in)	Rainfall Excess (in)
0	0	0	0
2	0.022	0.08	0
4	0.048	0.17	0.01
6	0.080	0.29	0.02
8	0.120	0.43	0.18
10	0.181	0.65	0.35
12	0.663	2.37	1.95
14	0.820	2.93	2.55
16	0.880	3.14	2.72
18	0.916	3.27	2.86
20	0.952	3.40	3.02
22	0.976	3.48	3.06
24	1.000	3.57	3.10

In order to formulate the Snyder unit graph use $C_p = 0.63$ and $C_t = 2.0$, values for Appalachians.

Eq. 2-37, $t_p = 0.95(2.0)[9(4.5)] + 0.74(2) = 7.2$ hrs, Use 7 hrs for ease in computing.
$T = 4\,t_p = 28.8$ hrs. , Use 29 hrs., We use this value of T since the basin is small and Eq. 2-39 would give a value of T much too large.

Eq. 2-38, $q_p = 640(0.63)/7.2 = 56$ cfs/sq mi
$Q_p = 56(200) = 11200$ cfs

From Fig. 2-33 obtain: W50 = 10 and W75 = 6 for a $q_p = 56$ cfs/ sq mi.
Plot the following trial unit graph.

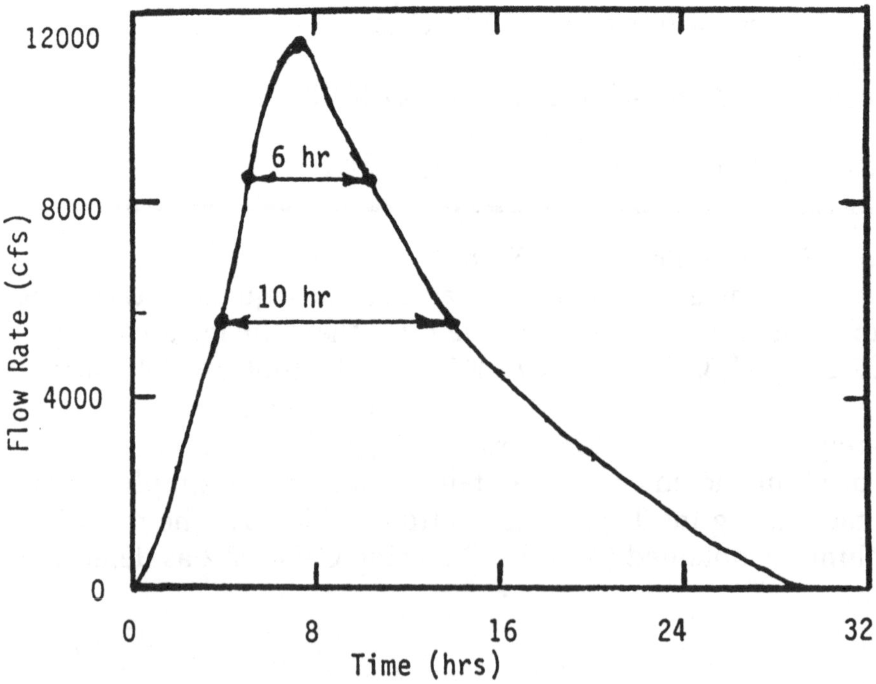

Prob. 2-42 Cont'd.

Read the following ordinates from the trial unit graph:

Time	Q (cfs)	Volume (cfs-hrs)	Corrected Q (cfs)
0	0		0
2	3000	3000	2890
4	5600	8600	5400
6	10200	15800	9840
7	11200	10600	10800
8	11000	11100	10600
10	8800	19800	8490
12	7000	15800	6750
14	5600	12600	5400
16	4500	10100	4340
18	3500	8000	3380
20	2800	6300	2710
22	2000	4800	1930
24	1300	3300	1250
26	800	2100	770
28	0	800	0
	Total =	132700 cfs-hrs	

Check the unit graph to see if it has a volume equal to 1 inch depth over the basin. Unit volume = (1/12)200(640) = 10667 AF = 12(10667) = 128004 cfs-hrs Thus, the unit graph has too much volume. The ordinates of the trial unit graph must be corrected in order to give a unit volume:

Corr. Factor = 128004/132700 = 0.96 The last column of the table has been obtained by multiplying column 2 by the correction factor 0.96. These corrected ordinates can now be used to generate the streamflow for the storm increments. The following contains the results.

Time (hrs)	Runoff for each 2-hr increment of precip from the first table (cfs)											Total Flow (cfs)
0												0
2	0											0
4	29	0										29
6	54	29	0									83
8	98	54	462	0								614
10	106	98	864	491	0							1559
12	85	106	1574	918	4624	0						7307
14	68	85	1698	1673	8640	1730	0					13894
16	54	68	1358	1804	15744	3240	491	0				22759
18	43	54	1080	1443	16976	5904	918	404	0			26822
20	34	43	864	1148	13584	6366	1673	756	462	0		24930
22	27	34	694	918	10800	5094	1804	1378	864	116	0	21729
24	19	27	541	738	8640	4050	1443	1485	1574	216	116	18849

Prob. 2-42 Cont'd.

26	12	19	433	575	6944	3240	1148	1189	1698	394	216	15868
28	8	12	309	461	5408	2604	918	945	1358	424	394	12841
30	0	8	200	328	4336	2028	738	756	1080	340	424	10238
32		0	123	212	3088	1626	575	608	864	270	340	7706
34			0	131	2000	1158	461	473	694	216	270	5403
36				0	1232	750	328	379	541	174	216	3620
38					0	462	212	270	433	135	174	1686
40						0	131	175	309	108	135	858
42							0	108	200	77	108	493
44								0	123	50	77	250
46									0	31	50	81
48										0	31	31
50											0	0

In the preceding table, the interior columns have been obtained by multiplying the unit graph ordinates by the precipitation. Columns 2 and 3 were obtained by multiplying the unit graph ordinates by 0.01 inches and column 4 was obtained by using 0.16, etc. Each successive column is lagged 2 hours later to account for the position of the rainfall increment in the storm. The last column is the composite hydrograph and is obtained the ordinates of all of the interior hydrographs horizontally so that the composite hydrograph ordinate is the sum of all the interior hydrographs summed for each time increment.

2-43 Use a time increment of 2 hours in Eq. 2-36 to analyze the beginning and end of runoff. $Q(t + \Delta t) = K^2 Q(t)$. Thus, $K = (Q(t + \Delta t)/Q(t))^{1/2}$. Values of K, calculated with this equation, are listed in column 3 of the following table.

Time (hrs)	Q (cfs)	K	Baseflow (cfs)	Runoff (cfs)	Vol. of Runoff (cfs-hrs)	Unit Graph Ordinate (cfs)
8	54		54	0	0	0
10	54	1.0	54	0	0	0
12	53	0.991	53	0	0	0
14	53	1.000	53	0	0	0
16	52	0.991	52	0	0	0
18	58	1.060	52	6	6	40
20	66	1.070	53	13	19	87
22	76	1.070	54	22	35	147
24	88		55	33	55	221
26	108		55	53	86	355
28	138		56	82	135	549
30	178		56	122	204	817
32	208		57	151	273	1012
34	233		58	175	326	1172
36	254		58	196	371	1313
38	251		59	192	388	1286
40	240		59	181	373	1213
42	230		60	170	351	1139

Prob. 2-43 Cont'd.

44	218		61	157	327	1052
46	200		62	138	295	925
48	190		62	128	266	858
50	177		63	114	242	764
52	157		64	93	207	623
54	142		64	78	171	522
56	130		65	65	143	435
58	120		66	54	119	362
60	111		66	45	99	302
62	103		67	36	81	241
64	96		67	29	65	194
66	90		68	22	51	147
68	86	0.95	69	17	39	114
70	82	0.976	69	13	30	87
72	79	0.981	70	9	22	60
74	76	0.981	70	6	15	40
76	74	0.981	71	3	9	20
78	72	0.986 ←	72	0	3	0
80	71	0.993	0	0	0	0
82	70	0.993				
84	69	0.993				
86	69	1.000				
88	68	0.993				
90	68	1.000				

Total Runoff Volume = 4806 cfs-hrs = 400 AF

Observing the calculated values of K, the runoff appears to start at 18 hours and recession appears to end at 78 hours. Baseflow has been proportioned linearly between these times: Example for22 hours:

$$\text{Baseflow} = 52 + (72 - 52)(22 - 16)/(78 - 16) = 54 \text{ cfs.}$$

Runoff has been calculated ad the difference between total flow(Col. 2) and baseflow (Col. 4).

The required volume for the unit graph is = 50(640)/12 = 2667 AF

To compute unit graph ordinates, the hydrograph ordinates in Col.1 are multiplied by the factor = 2667/400 = 6.7

The resulting unit graph ordinates are shown Col.7.

2-44. The following table has been developed to calculate the ordinates of the resulting hydrograph. Column 2 is the increments in rainfall excess as given. Columns 3,4,and 5 are the hydrograph ordinates as calculated by multiplying the increments in rainfall excess by the unit graph ordinates. Column 6 is the composite hydrograph obtained by adding the individual horizontal lines.

Time (hrs)	Incr. Rainfall Excess (in)	Runoff from 1.1 in. (cfs)	Runoff from 1.9 in. (cfs)	Runoff from 0.7 in. (cfs)	Total Runoff (cfs)
0	1.1	0	0	0	0
2		44			44
4		96			96
6		162			162
8		243			243
10		390			390
12		604			604
14		899			899
16		1113			1113
18		1289			1289
20-		1444			1444
22		1415			1415
24	1.9	1334	0		1334
26		1253	76		1329
28		1157	165		1322
30		1018	279		1297
32		944	420		1364
34		840	674		1514
36		685	1043		1728
38		574	1552		2126
40		478	1923		2401
42		398	2227		2625
44		332	2495		2827
46		265	2443		2708
48	0.7	213	2305	0	2518
50		162	2164	28	2354
52		125	1999	61	2185
54		96	1758	103	1957
56		66	1630	155	1851
58		44	1452	249	1745
60		22	1184	384	1590
62		0	992	572	1564
64			826	708	1534
66			688	820	1508
68			574	919	1493
70			458	900	1358
72			369	849	1218
74			279	797	1076
76			217	736	953

Prob. 2-44 Cont'd.

78	165	648	813
80	114	601	715
82	76	535	611
84	38	436	474
86	0	365	365
88		304	304
90		253	253
92		211	211
94		169	169
96		136	136
98		103	103
100		80	80
102		61	61
104		42	42
106		28	28
108		14	14
110		0	0

2-45. First plot the hydrograph as shown below:

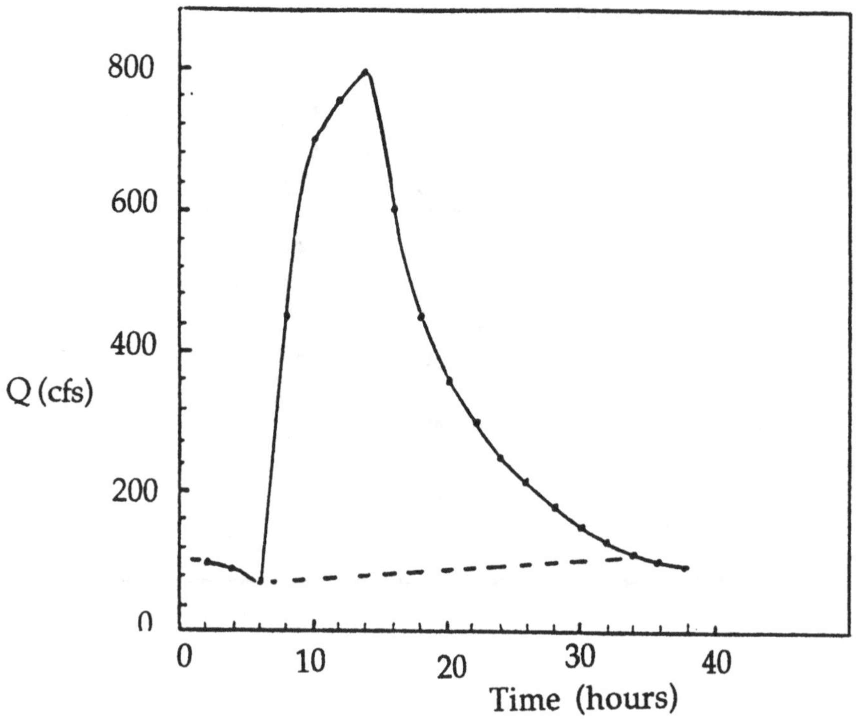

As shown on the hydrograph, the surface runoff undoubtedly starts at 6 hrs. Also, by analyzing the recession curve using Eq. 2-36, it will be indicated that the surface runoff probably ceases at 34 hours. Divide the surface runoff from base flow by a straight line as shown. The ordinates of the surface runoff hydrograph are determined and tabulated in the following table:

Time (hrs)	Surface R_o Q (cfs)	Unitgraph Ordinates Q (cfs)
0	0	0
2	375	783
4	620	1294
6	680	1420
8	725	1514
10	520	1086
12	360	752
14	270	564
16	205	428
18	155	324
20	118	246
22	80	167
24	45	94
26	20	42
28	0	0

Prob. 2-45 Cont'd.

Next the volume of surface runoff from column 2 (above) is calculated:

$$\forall = \sum Q \, \Delta t$$

$$= (375 + 620 + \ldots \qquad) \ ft^3/s \times 2 \ hr \times 3600 \ s/hr$$

$$= 3.00456 \times 10^7 \ ft^3$$

Next compute the depth of surface runoff

$d = V/A$ where $A = 27 \times 5{,}280^2 = 75.272 \times 10^7 \ ft^2$

$$d = 3.00456 \ ft^3 / 75.272 \times 10^7 \ ft^2$$

$= 0.0399$ ft = <u>0.479 in.</u>

With this depth of runoff (0.479 in.) the discharge in values in column 2 (above) are divided by 0.479 to obtain the unitgraph ordinates. These are shown in <u>column 3 of the above table</u>.

2-46. Compute the basic parameters

$$t_p = 0.95 \ C_t \ (LL_{CA})^{0.3} + 0.74 \ t_r \qquad \text{(Eq. 2-37)}$$

$$q_p = 640 \ C_p / t_p$$

$$T = 3 + t_p/8$$

where

t_r = 4 hrs (assumed)

L = 11 mi.

L_{CA} = 5.5 mi

$C_p = 0.63$, $C_t = 2.00$ (assumed, based on the table on page 81)

Thus, from Eqs. 2-37 and 2-38

$t_p = 0.95 \ (2.00) \ (11 \times 5.5)^{0.30} + (0.74 \times 4) =$ <u>9.47 hours</u>

$q_p = 640 \ (0.63/9.47) = 42.6$ cfs/square mile

and, $Q_p = 42.6 \times 27 =$ <u>1150 cfs</u>

Prob. 2-46 Cont'd.

$T = 3 + 9.47/8 = \underline{4.18 \text{ days}} = 100 \text{ hrs}$

or $T = 4t_p = 37.9 \text{ hrs} \approx 38 \text{ hrs}$ ← (use this--small basin)

Further refinement according to Fig. 2-33.

W-50 = 13 hrs.

W-75 = 18 hrs

Plot key parameters of synthetic unit hydrograph and complete the hydrograph by sketching in curve. (See trial unit hydrograph below.)

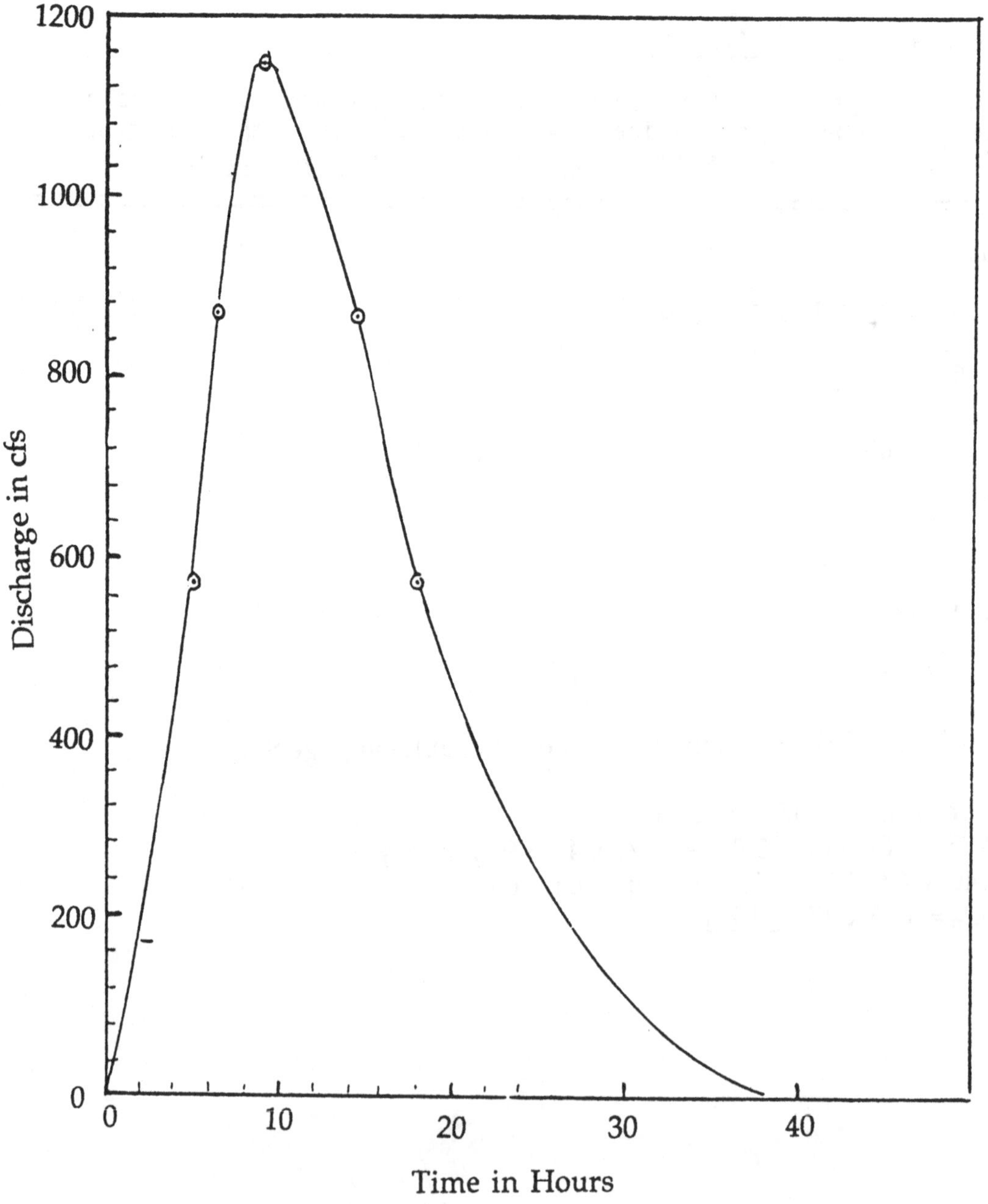

Prob. 2-46 Cont'd.

Now check the unit hydrograph as graphed to see if it indeed represents a runoff of 1 in. over the basin.

The 2 hr ordinates of the unit graph are given below (column 2).

Time (hrs)	Q (cfs)	Q of Unitgraph (cfs)
0	0	0
2	150	153
4	400	408
6	760	776
8	1080	1102
10	1140	1163
12	1030	1051
14	920	939
16	740	755
18	580	592
20	470	480
22	365	372
24	280	286
26	220	224
28	160	163
30	110	112
32	70	71
34	40	41
36	20	20
38	0	0

$$\forall = \Sigma Q \Delta t = (0 + 150 + 400 + \ldots) \times 2 \times 3600 = 6.145 \times 10^7 \text{ ft}^3$$

$$A = 27 \times 5{,}280^2 = 7.527 \times 10^8 \text{ ft}^2$$

$$\text{Depth of runoff} = 6.145 \times 10^7 \text{ ft}^3 / 7.527 \times 10^8 \text{ ft}^2$$

$$= 0.0816 \text{ ft}$$

$$= 0.980 \text{ in.}$$

Since the runoff is 0.980 in. rather than the desired 1.00 in., divide all the ordinates of the trial unitgraph by 0.980. The resulting ordinates are given in column 3 of the above table.

2-47. Drainage Area = 27 mi2

t_c = 4 hours

Eq. 2-40 $t_p = (t_r/2) + 0.6\, t_c = (2/2) + 0.6(4) = 3.4$ hours

Eq. 2-41 $Q_p = 484\,(A/t_p) = 484\,(27/3.4) = 3844$ cfs

The increment in time used in the unit hydrograph should be $\leq t_r/5 = 0.8$ hours.

Using the values of t_p and Q_p calculated above and the tabulated values shown in Table 2-13 the following unit-hydrograph ordinates are developed:

Time (hours)	Q (cfs)	Time (hours)	Q (cfs)	Time (hours)	Q (cfs)
0	0*	3.7	3806	8.2	565*
0.3	115	4.1	3575*	8.8	411*
0.7	384*	4.4	3306	9.5	296*
1.1	730	4.8	2998*	10.2	211*
1.4	1192*	5.1	2614	10.9	154*
1.7	1807	5.4	2153*	11.6	112*
2.0	2537*	5.8	1768	12.2	81*
2.4	3152	6.1	1499*	12.9	58*
2.7	3575*	6.5	1268	13.6	42*
3.1	3806	6.8	1076*	15.3	19*
3.4	3844*	7.5	796*	17.0	0*

Those unit hydrograph ordinates indicated by an asterisk * are the only ones used in the further computations since they are sufficient for the accuracy desired.
The following table shows the hydrograph computations. The first two columns, the time and the Accumulative Rainfall Excess are taken directly from the solution for Prob. 2-38. The 3rd column is the incremental precipitation computed from column 2. The next 10 columns are the hydrograph ordinates produced by multiplying the unit asterisked unit hydrograph ordinates from the preceding table by the precipitation excess in column 3.

Prob. 2-47 Cont'd.

T hrs.	P in.	Inc. In.											Flow cfs
0	0												0
2	0												0
4	0												0
6	.02	.02	0										0
8	.85	.83	8	0									8
10	1.2	.35	24	318	0								342
12	1.24	.04	51	989	134	0							974
14	1.25	.01	72	2106	417	15	0						2610
16	1.28	.03	77	2967	888	48	4	0					3984
18	1.30	.02	72	3190	1251	101	12	12	0				4638
20	1.32	.02	60	2967	1345	143	25	36	8	0			4584
22	1.36	.04	43	2488	1251	154	36	76	24	8	0		4080
24	1.40	.04	30	1787	1049	143	38	147	51	4	15	0	3244
26			21	1244	754	120	36	115	72	51	48	15	2476
28			16	894	525	86	30	107	77	72	101	48	1956
30			11	661	377	60	22	89	72	77	143	101	1613
32			8	469	279	43	15	64	60	72	154	143	1307
34			6	341	198	32	11	45	43	60	143	154	1033
36			4	246	144	23	8	32	30	43	120	143	793
38			3	175	104	16	6	24	21	30	86	120	685
40			2	128	74	11	4	17	16	21	60	86	419
42			2	93	54	8	3	12	11	16	43	60	210
44			1	67	39	6	2	9	8	11	32	43	208
46			1	48	28	4	2	6	6	8	23	32	158
48			0	35	20	3	1	5	4	6	16	23	113
50				16	15	2	1	3	3	4	11	16	71
52				0	7	2	0	2	2	3	8	11	35
54					0	1	0	2	2	2	6	8	21
56						0	0	1	1	2	4	6	14
58							0	0	1	1	3	4	9
60									0	1	2	3	6
62										0	1	2	3
64											0	1	1
66												0	0

2-48. The sketch below shows how the baseflow was separated from surface runoff.

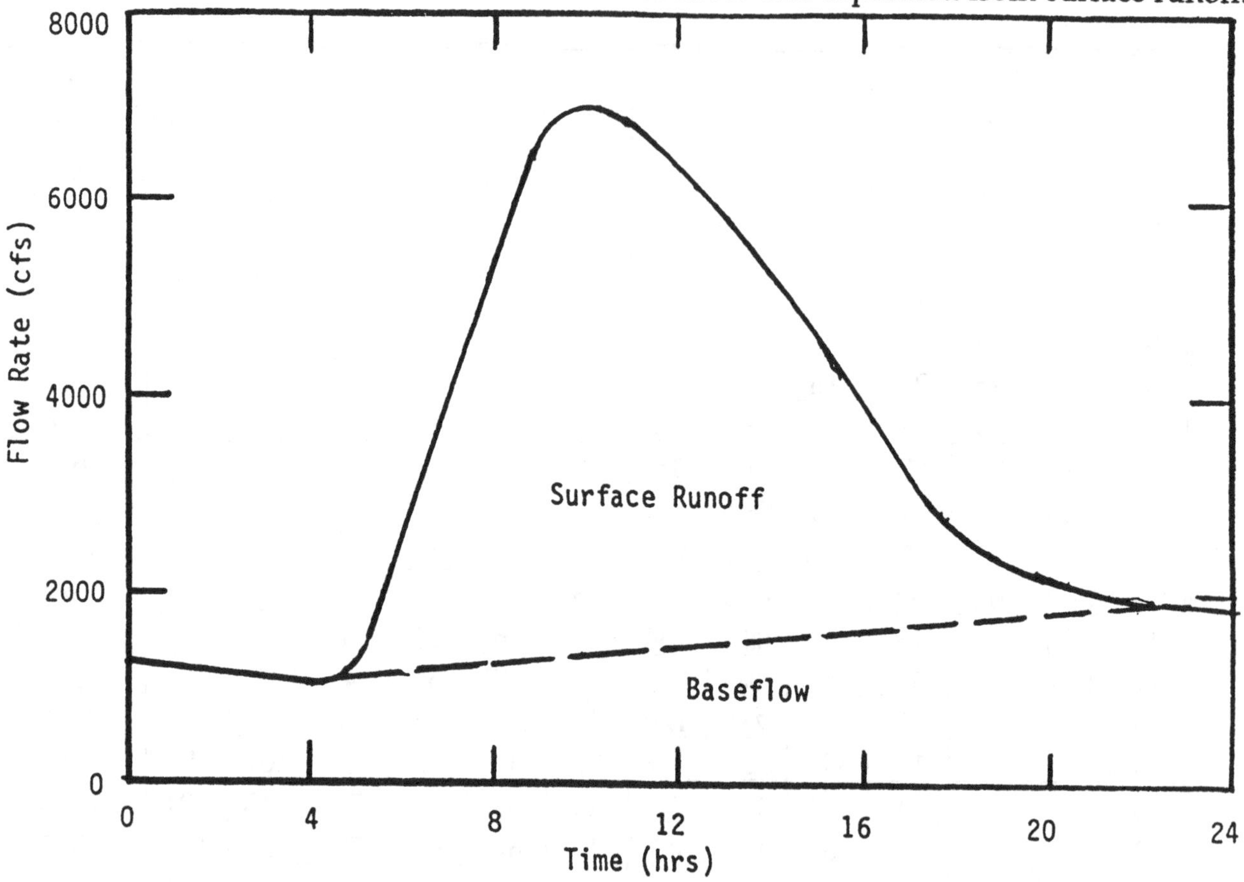

The hydrograph and baseflow ordinates were read from the preceding graph.

Time (hrs)	Q (cfs)	Baseflow (cfs)	Surface Runoff (cfs)	Runoff Volume (cfs-hrs)	Baseflow Volume (cfs-hrs)	Unit Graph Ordinate (cfs)
0	1300	1300	0	0	0	0
4	1000	1000	0	0	0	0
6	2500	1100	1400	1400	2100	233
8	5400	1200	4200	5600	2300	701
10	7000	1300	5700	9900	2500	952
12	6400	1400	5000	10700	2700	835
14	5300	1500	3800	8800	2900	635
16	4000	1600	2400	6200	3100	401
18	2700	1700	1000	3400	3300	167
20	2200	1800	400	1400	3500	67
22	1900	1900	0	400	3700	0
		Totals =		47800	26100	

Prob. 2-48 Con't.

The runoff volume= 47800/12 = 3983 AF
The baseflow volume = 26100/12 = 2175 AF

a. Depth of runoff = 3983(12)/8000 = <u>5.97 inches</u>

b. Required Volume of Unit Graph = 1.0 (8000)/12 = 667 AF
Factor for Unit Graph ordinates = 667/3983 = 0.167
The surface runoff ordinates have been multiplied by this factor to obtain the unit graph ordinates in Column 7. The resulting unit graph is plotted in the following graph.

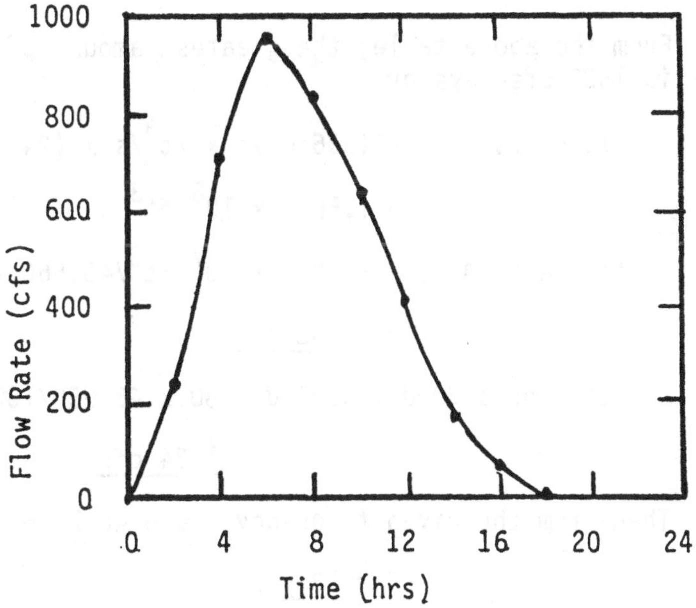

2-49. The farthest point is $[(500)(500)2]^{1/2} = 707$ feet from A.

The time of flow from the fartherst point to A = 707/0.5 - 1414 seconds = 23.6 min.
The peak intensity resd from graph for 23.6 minutees is 1.4 in/hr.
Area = 500(500)/43560 = 5.67 acres.
From Table 2-9, C = 0.70 to 0.95. Use average value = 0.82
Thus, Q_p = CIA = 0.82(1.4)5.7 6.5 acre-inches/hour = <u>6.5 cfs</u>

2-50 For a six-year drought ($T_r = 6$) the following table can be developed from the given low-flow frequency graph. The values for Avg. Flow in column 2 are read directly from the graph.

Duration (days)	Avg. flow (cfs)	Inflow volume (cfs-days)	Demand volume (cfs-days)	Inflow-demand (cfs-days)
1	0.51	0.51	25	-24.5
7	0.89	6.23	175	-168.8
20	1.7	34.0	500	-466.0
30	2.8	84.0	750	-666.0
60	4.8	288.0	1500	-1212.0
120	9.7	1164.0	3000	-1836.0
183	20.1	3678.0	4575	-897.0

a) From the above table, the greatest amount of storage needed is 1836 cfs-days or

$$1836 \text{ cfs-days} = 1836 \text{ days} \times ft^3/s \times (24 \times 3600s)day$$

$$= 1.5836 \times 10^8 \; ft^3$$

$$\text{Storage is A.F.} = 15.863 \times 10^7 \; ft^3/43{,}560 \; ft^3/A.F.$$

$$= \underline{3642 \text{ A.F.}}$$

b) 80 A.F. for a 30 day period $= 80 \times 43{,}650/(30 \times 24 \times 3600)$

$$= \underline{1.34 \text{ cfs}}$$

Then from the given frequency graph we have

$$T \approx \underline{13 \text{ yrs}}$$

CHAPTER THREE

3-1. Eq. (3-2) Vol. water = $n V_t$ = 0.48 (100) = 48m^3.

3-2. Total weight = 85 lbs

Wt. water drained = 85-73 = 12 lbs Vol = 12/62.4 = 0.19 ft^3

Total water = 85-51 = 34 lbs Vol = 34/62.4 = 0.54 ft^3

Vol. voids = Vol. of water = 34/62.4 = 0.54 ft^3

Eq. 3-1	n	= V_v/V_f	= 0.54/1.0 = <u>0.54</u>	
Eq. 3-4	S_y	= 0.19/1.0	= <u>0.19</u>	
Eq. 3-5	S_r	= (1-0.19)/1 = <u>0.81</u>		
Unit Wt.	Solids	= 51/1-0.54	= 51/0.46 = 110.9 lb/ft^3	
	S_g	= 110.9/62.4 = <u>1.78</u>		

3.3 From Table 3-1; n = 0.11, Sy = 0.06

Vol aquifer including water = 4500(1500) (640) = 4,320,000,000 AF

Total Vol. Water = 4,320,000,000 (0.11) = 475,200,000 AF

Vol that could be withdrawn = 4,320,000,000 (.06) = <u>259,200,000 AF</u>

3.4 L = 2(5280/3.28 = 3219 m

gradient = 100/(5280)(2) = 0.0095

From Table 3-1

Problem 3-4 Cont'd.

a. K(Gravel) = 7000 m/day

Eq 3-11) V = 7000 (.0095) = 66.5 m/day

Time = L/V = 3219/166.5

= <u>48 days</u>

b. K (Clay) = 0.0001 m/day

V = 0.0001 (.0095) = 0.00000095 m/day

T = 3219/0.00000095

= 3388 million days

= <u>9.3 million years</u>

c. K (Sand) - 0.5 m/day

V = 0.5(.0095):

= 0.0048 m/day

T = 3219/.0048

= 670625 days

= <u>1837 years</u>

3.5 Gradient = 100/(2)(5280) = 0.0095

From Table 3-1. K = 1000 m/day L = 2(5280)/3.28 = 3219 m

Eq. 3-11) V = 1000(0.0095) = 9.5 m/day

T = 3219/9.5 = <u>338 days</u>

3.6 For all three tests:

$A = \Pi(4)^2 4 = 12.56$ sq. Inches

$L = 5(12) = 60$ in.

$Q = 0.227/7.48 = 0.0303\ ft^3/hr = 52.36\ in^3/hr$

$V = Q/A = 52.36/12.56 = 4.17$ in/hr

a)	b)	c)
$h_1 = 185$	$h_1 = 77$	$h_1 = 39$
$h_2 = 34$	$h_2 = 35$	$h_2 = 36$
$h_1 - h_2 = 151$	$h_1 - h_2 = 42$	$h_1 - h_2 = 3$

	a)	b)	c)
$(h_1-h_2)/L$	= 151/60 = 2.517	= 42/60 = 0.7	= 3/60 = 0.05
V	= 4.17	= 4.17	= 4.17
$K = V/(h_1-h_2)/L]$	= 1.66 in/hr	= 5.96 in/hr	= 83.4 in/hr
	= 3.32 ft/day	= 11.92 ft/day	= 166.8 ft/day
	= <u>1.01 m/day</u>	= <u>3.63 m/day</u>	= <u>60.9 m/day</u>

3.7 $n = 0.30$ $K = 10$ ft/day

Area pipe = $12.56\ in^2$ Area Voids = $12.56(.3) = 3.77\ in^2$

Eq. 3-10. $V = K\ (h_2-h_1)/L = -10(-0.005) = 0.05$ ft/day

V (in voids) = 0.05 (12.56)/3.77 = <u>0.17 ft/day</u>

3.8 A (horiz. tube) $= \Pi (10)^2/4 = 78.54$ sq. cm.
A (vert. tube) $= \Pi (40)^2/4 = 1256.6$ sq. cm.

Vol. Water = L(A) = (200-180) (1256.6) = 25132 cm^3
V = 25132/(78.54)/4 = 79.997 cm/hr.
K avg = 79.997/[(190-15)/100] = 45.7 cm/hr = 0.457 m/hr
K min = 79.997/[(200-15)/100] = 43.2 cm/hr = 0.432 m/hr
K max = 79.997/[(180-15)/100] = 48.5 cm/hr = 0.485 m/hr

3-9. n = 0.5, $(h_2-h_1)/L$ = -100/100 = -1
$A = \Pi (10)^2/4 = 78.5$ sq. cm.
V = 2000/78.5/10 = 2.55 cm/min = 120 ft/day

From Fig. 3-4, n = 0.47

Yes, a decrease of [(0.5 - 0.47)/0.5] 100 = 6%

3-10. B = 10, K = 0.06 m/day, h_1 = 7 m, h_2 = 10/3.28 = 3.05 m,
L = 300 m,

Eq. 3-15 q = -(0.06)(3.05-7)(10)/300 = 0.0079 m^3/day
Q = 0.0079(100) = 0.79 m^3/day

3-11. B = 19′, L = 400′, h_1-h_2 = 10′, Q = 10 gpm

Eq. 3-15 q = $-kB(h_2-h_1)/L$
Q = 10/(7.48) = 1.34 ft^3/min
q = Q/200 = 1.34/200 = 0.0067 ft^3/min/ft.

Problem 3-11 Cont'd.

$K = qL/[B(h_2-h_1)] = 0.0134(400)/[19(10)]$
$= 0.028$ ft/min
$=$ <u>40.6 ft/day</u>

3-12. $B = 20$, $h_1 = 30$, $h_2 = 6$, $L = 300$

Eq. 3-22: $q = (K/2L)[2B(h_1-B) + (B^2-h_2^2)$

From Table 3-1 $K = 0.06$ to 120m/day, use $K = 60$m/day
$= 197$ ft/day

$q = [197/2(300)][2(20)(30-20)+(20^2-6^2)] = 250$ ft^3/day
$Q = 100(250) = 25000$ ft^3/day $=$ <u>0.289 cfs</u>

Eq. 3-15: $L_1 = KB(h_1 - B)/q = 197(20)(30-20)/250 =$ <u>158 ft</u>

3-13. $k = 0.1$ ft/day, $h_1 = 20$, $Z_2-Z_1 = 30$
$L = (30^2 + 600^2)^{1/2} = 600.7$ ft: use 600 ft., $B=20$ $h_2 = 40$
$h_1 + (Z_1-Z_2) = 20 + 30 = 50 > h_2 = 40$, water will flow out

Eq. 3-23: $q = -kB(h_2-h_1 + Z_2-Z_1)/L$
$= -0.1(20)(40-20 - 30)/600$

$q =$ <u>0.033 ft^3/day</u>

3-14. r_w = 10/2 = 5 in = 0.42 ft., h_0 = 500 ft, K = 0.25 ft/day
Q = 100 gpm = 19251 ft^3/day

Eq. 3-26: $Q = \Pi K(h_2^2 - h_1^2)/\ln(r_2/r_1)$

$19251 = \Pi(0.25)(500^2 - h_w^2)/\ln(1000/0.42)$

$500^2 - h_w^2 = 24511 \ln(100/,42) = 190,579$
$h_w^2 = 250,000 - 190,579 = 59,421$
h_w = <u>244 feet</u>

$h_2^2 - h_1^2 = Q \ln(r_2/r_1)/\Pi K = 19,251 \ln(1000/250)/0.25 \Pi = 33,980$
$h_1^2 = (500)^2 - 33,980\ 250,000 - 33,980 = 216,020$
$h_1 = 465$ feet
Drawdown = 500 - 465 = <u>35 feet</u>

3-15. Q = 100 gpm = 19251 ft^3/day, r_1 = 0.5, h_1 = 300 - 110 = 190 ft
r_2 = 400, h_2 = 300 - 5 = 295 ft.

From Eq. 3-26: $K = Q \ln(r_1/r_2)/[\Pi (h_2^2 - h_1^2)]$
$= 19,251 \ln(400/0.5)/[(295^2 - 190^2)\Pi]$
= <u>0.804 feet per day</u>

3-16. rw = 0:25 m, K = 0.8 m/day, h_0 = 330, B = 30 m,
Q = 0.8 m^3/min = 1152 m^3/day, r_2 = 1000 ft = 304.8 m

From Eq. 3-29, $h_2 - h_1 = Q \ln(r_2/r_1)/2\Pi KB$
$= 1,152 \ln(304.8/0.25)/2\Pi(0.8)(30)$
= 54.3 meters

Problem 3-16 Cont'd. $hw = 330 - 54.3 = \underline{275.7\ m}$

$$330 - h_1 = 1{,}152 \ln(304.8/50)/2\Pi(0.8)(30)$$
$$= 13.8$$
$$h_1 = 330 - 13.8 = 316.2 \text{ meters}$$
$$\text{height above aquifer} = 316.2 - 30 = \underline{286.2 \text{ meters}}$$

3-17. $B = 85$ ft, $r_W = 43$ in $= 3.58$ ft., $K = 10$ ft/day, $h_0 = 170$ ft., assume radius of reference $= 1000$ ft.

Eq. 3-29: $Q = 2\Pi K(h_2 - h_1)/\ln(r_2/r_1)$

$$= 2\Pi(10)(85)(170\text{-}85)/\ln(1000/3.58)$$
$$= \underline{8{,}0598 \text{ cubic feet per day}}$$

at $r = 20$ ft., $B = 85$ ft., $A = 85\Pi(2)(20) = 10{,}681$ ft.

$$V_{avg} = 80{,}598/10{,}681 = \underline{7.5 \text{ ft/day}}$$

3-18. $r = 70$, $Q = 150$ gpm $= 20.0$ ft^3/min

Time (hrs)	Time (min)	Drawdown (ft)	r^2/t	u	W(u)
0	0	0			
0.02	1.2	0.1	4083	0.2	1.22
0.03	1.8	0.2	2722	0.4	0.70
0.04	2.4	0.3	2042	0.6	0.45
0.06	3.6	0.5	1361	0.8	0.31
0.11	6.6	0.8	742	1.0	0.219
0.25	15.0	1.3	327	2.0	0.049
0.43	25.8	1.6	189	4.0	0.0038
1.00	60.0	2.2	82		
3.90	234.0	3.0	21		

Problem 3-18 Cont'd.

U vs W(u) and drawdown vs r^2/t are plotted in the accompanying figure. A match point is chosen (see figure). Values of parameters at the match point are as follows:

$h_0 - h = 0.5$, $r^2/t = 1361$, $u = 0.53$, $W(u) = 0.50$

Eq. 3-36: $KB = (Q/4\pi)(W(u))/(h_0 - h) = [20/4\pi]0.50/0.5 = \underline{1.59\ ft^2/min}$

Eq. 3-37: $S = 4KB[u/(r^2/t)] = 4(1.59)(0.53/1361) = \underline{0.0025}$

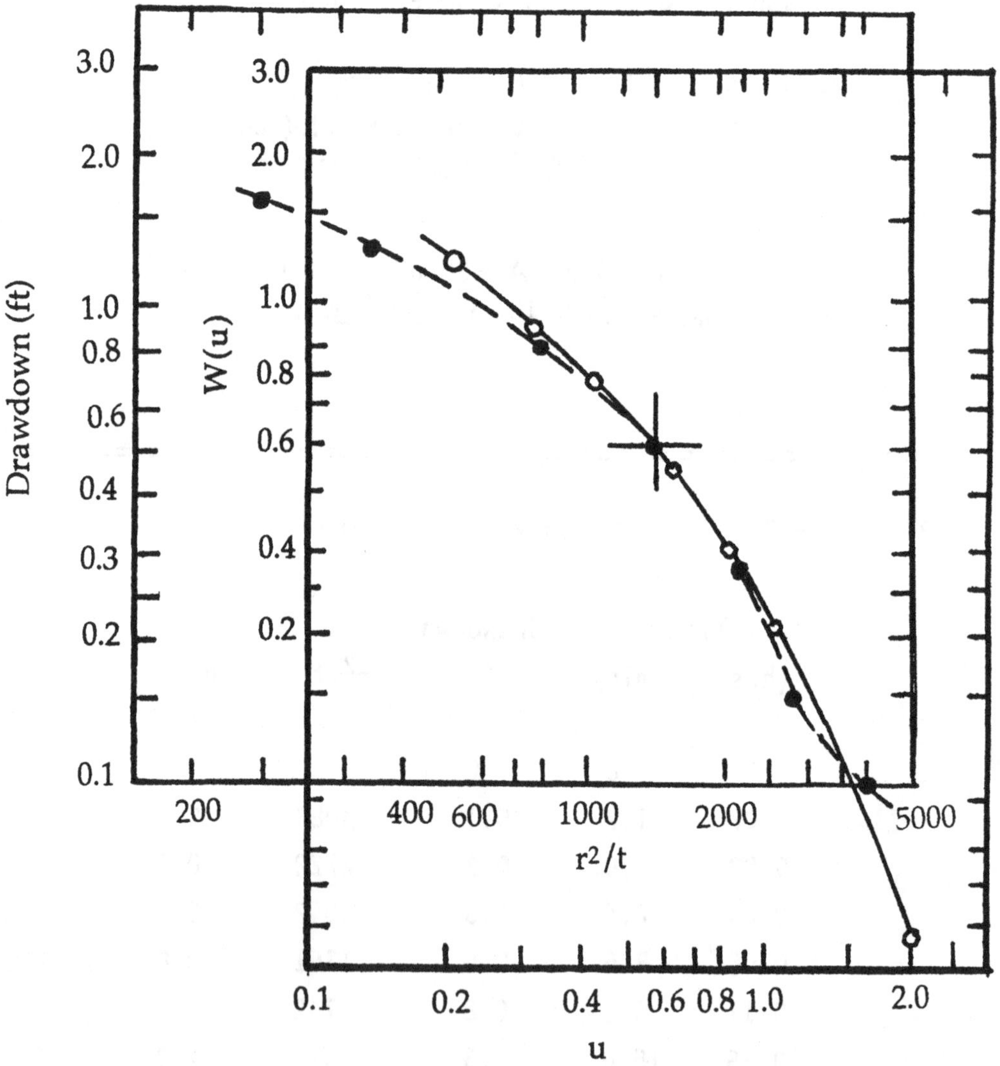

3-19. Q = 0.41 cfs = 24.6 cfm

Time (hrs)	Drawdown (r = 320) (ft)	(r = 640) (ft)	h (640) - h (320) (ft)
0.024	0.27	0.06	0.21
0.120	0.54	0.32	0.22
0.240	0.81	0.48	0.33
1.200	1.23	0.88	0.35
2.4	1.41	1.06	.35
12.0	1.83	1.47	.36
24.0	2.01	1.65	.36
120.0	2.43	2.07	.36
240.0	2.61	2.25	.36

Note that h(640) - h(320) = Const. = 0.36 for last four times.

Thus, we can use Eq. 3-29.

Eq. 3-29: $KB = Q \ln(r_2/r_1)/ 2\Pi(h_2 - h_1) = 0.41 \ln)640/320)/2\Pi(0.36)$
$= 0.13 \text{ ft}^2/\text{sec.} = \underline{7.8 \text{ ft}^2/\text{min.}}$

Eq.3-30: $W = 4\Pi(h_0 - h) KB/Q = 4\Pi(2.61)7.8/24.6 = 10.39$

From Table 3-2, u = 0.00001773

Eq. 3-32: $S = u(4KBT)/r^2 = 0.00001773(4)7.8(14{,}400)/(320^2) = \underline{0.00008}$

3-20. $Q = 0.41$ cfs $= 24.6$ cfm

Time (hrs)	$(h_0 - h)$ r = 320 (ft)	$(h_0 - h)$ r = 640 (ft)
0.024	0.27	0.06
0.120	0.54	0.32
0.240	0.81	0.48
1.20	1.23	0.88
2.4	1.41	1.06
12.0	1.83	1.47
24.0	2.01	1.65
120.0	2.43	2.07
240.0	2.61	2.25

Eq. 3-37: $4\pi KB(h_0 - h)/Q = \ln(2.25\ kBt/r^2S)$

We get simultaneous equations for each value of t

For $t = 240,\ r = 320,\ h_0 - h = 2.61$

$$4\pi KB(2.61)/24.61 = \ln[2.25KB(14{,}400)/(320)^2\ S]$$

$$1.33\ KB = -1.151 + \ln(KB/S) \qquad (1)$$

For $t = 240, r = 640, h_0 - h = 2.25$

$$4\pi KB(2.25)/24.61 = \ln[2.25KB(14{,}400)/(640)^2 S]$$

$$1.149\ KB = -2.537 + \ln(KB/S) \qquad (2)$$

Subtracting (2) from (1)

$(1.33-1.149)KB = -1.151 + 2.537 = 1.386$ $\underline{KB = 7.66\ ft^2/min}$

Problem 3-20 Cont'd.

From (2) $\ln(KB/S) = 1.149(7.66) + 2.587 = 11.34$

$KB/S = 84120$

$S = 7.6/84120 = \underline{0.00009}$

3-21. $Q = 100$ gpm $= 100/7.48 = 13.37$ ft^3/min, $r_w = 1$ ft

Assume a radius of influence of 1000 ft

From Eq. 3-38) $K = Q\ln(r/r_w)/(h_w^2 - h^2)\pi$

$= 13.37 \ln(1000/1)/[(115)^2 - (100)^2]$

$= 0.0091$ ft/min $= 13.13$ ft/day

$= \underline{4.00 \text{ m/day}}$

Eq. 3-38: $h = [(115)^2 - 13.37 \ln(80/1)/(0.286)]^{1/2}$

$= \underline{105.7 \text{ ft}}$

3-22. $Q = 30$ gpm $= 4.01$ cfm, $r_w = 1$ ft

$K = Q\ln(r/r_w)/(h_w^2 - h^2)\pi$

$= 4.01 \ln(600/1)/[(230)^2 - (200)^2]\pi$

$= 0.00063$ ft./min

Now integrate under the water table. This can be done readily numerically.

$$V = \int_0^{600} 2\Pi r(h - h_0)dr$$

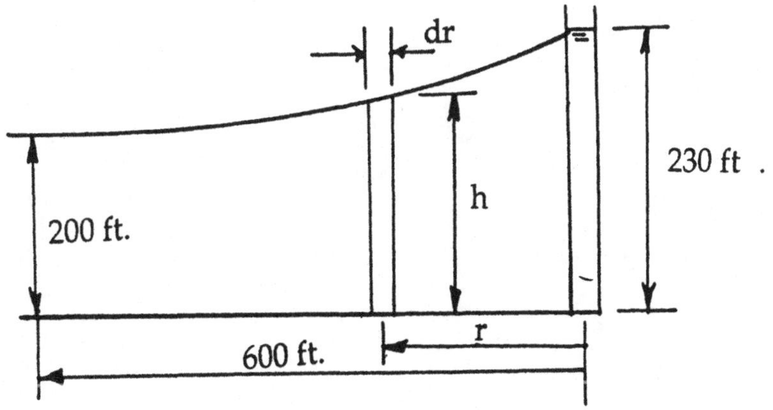

Problem 3-22 Cont'd.

Calculate h for each r in the following table. Use Eq. 3-38

r	h	$(h-h_0)$	$2\Pi r(h - h_0)$	$[2\Pi r(h - h_0)]$ Avg.	r	V
1	230	30	188			
2	226.9	26.9	338	263	1	263
3	225.2	25.2	475	406	1	406
4	223.9	23.9	601	538	1	538
5	222.9	22.9	719	660	1	660
7	221.4	21.4	941	830	2	1660
10	219.7	19.7	1238	1090	3	3270
15	217.9	17.9	1687	1462	5	7310
20	216.5	16.5	2073	1880	5	9400
50	212.3	12.3	3864	2968	30	89040
100	209.0	9.0	5655	4760	50	238000
150	207.0	7.0	6597	6126	50	306300
200	205.6	5.6	7037	6817	50	340850
300	203.6	3.6	6786	6911	100	691100
400	202.2	2.2	5529	6158	100	615800
500	201.1	1.1	3456	4492	100	449200
600	200.0	0	0	1728	100	172800

Total Volume = 2,926,597 ft^3

3-23. r_w = 30 in = 1.5 ft, K = 40 ft/day, Q = 400 gpm = 53.5 ft^3/min
h_o = 200 ft, R = 300 ft = 77005 ft^3/day

We will use a negative image well as shown in Fig. 3-13. The co-ordinates are calculated using Eq. 3-27 as shown in the following table.

From Eq. 3-27: $h = [h_2^2 - Q \ln (r_2/r)/\pi K]^{1/2}$

For the real well: $h = [(200)^2 - 77{,}005 \ln(1000/r)/40\pi]^{1/2}$
$= [(200)^2 - 612.8 \ln(1000/r)]^{1/2}$

For the image well: $r_i = 600 - r$

The following table is completed using the above equations.

Real Well		Image Well			
r (ft)	h (ft)	r (ft)	h (ft)	h_0-h (ft)	Combined Drawdown (ft)
1.5	189.8	598.5	199.2	0.8	189.8 + 0.8 = 190.6
3.0	190.8	597	199.2	0.8	190.8 + 0.8 = 191.6
10.0	192.8	590	199.2	0.8	192.8 + 0.8 = 193.6
100.0	196.4	500	198.9	1.1	196.4 + 1.1 = 197.5
200.0	197.5	400	198.6	1.4	197.5 + 1.4 = 198.9
300.0	198.1	300	198.1	1.9	198.1 + 1.9 = 200.0

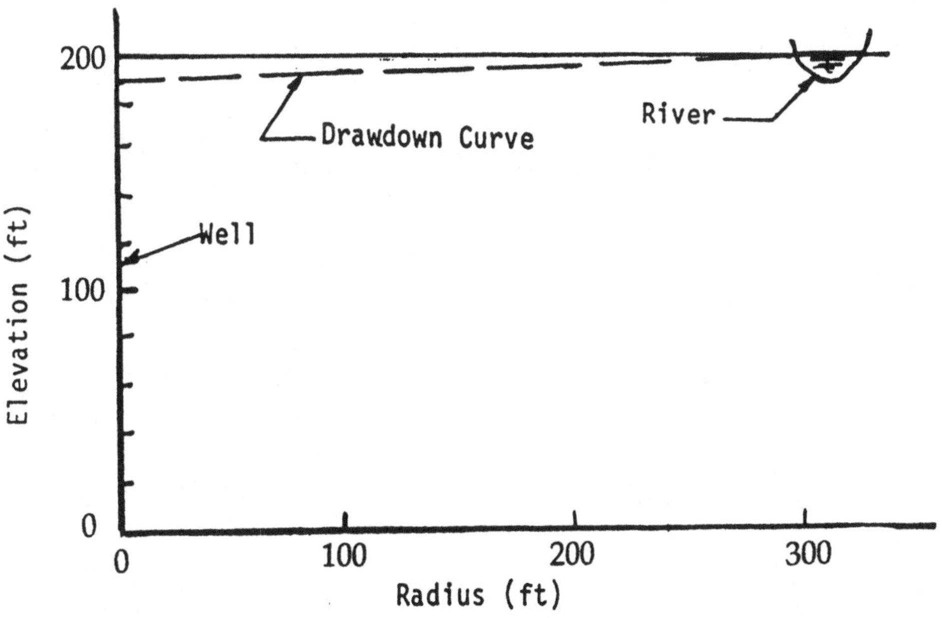

3-24. $Q = 400$ gpm $= 77{,}005$ ft^3/day, $K = 40$ ft/day

$$h = [h_2^2 - Q \ln(r_2/r)/\pi K]^{1/2}$$
$$= [(200)^2 - 77{,}005 \ln(1000/r)/\pi 40]^{1/2}$$
$$= [(200)^2 - 612.8 \ln(1000/r)]^{1/2}$$

a) Real Well

$r = 150$ ft
$h = 197.1$ ft

Image Well

$r = 450$
$h = 198.7$
$h_0 - h = 1.3$

Actual GW surface is 197.1 + 1.3 = <u>198.4 ft</u> above bottom of aquifer

b) $r = 150$ ft
$h = 197.1$ ft

$r = 750$
$h = 199.6$
$h_0 - h = 0.4$

Actual GW surface is 197.1 + 0.4 = <u>197.5 ft</u> above bottom of aquifer

3-25. $Q = 400$ gpm = 77005 ft^3/day, $K = 40$ ft/day

$$h = [h_2^2 - Q\ln(r_2/r)/\pi K]^{1/2} = [(200)^2 - 77{,}005\ln(1000/r)/40\pi]^{1/2}$$
$$= [(200)^2 - 612.8\ln(1000/r)]^{1/2}$$

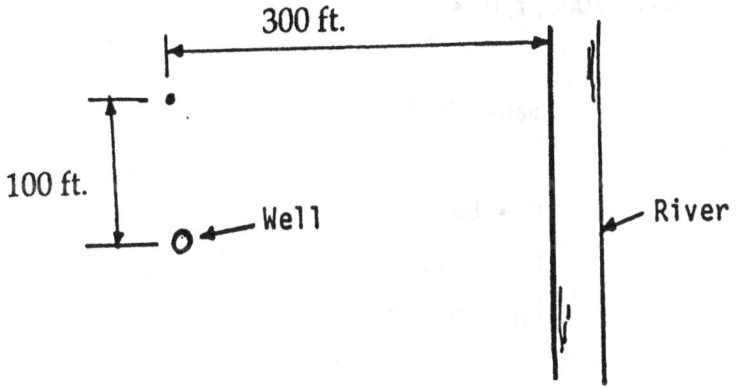

Real Well	Image Well
$r = 100$	$r = [(100)^2] + (600)^2] = 608.3$ ft
$h = 196.4$	$h = 199.2$
	$h_o - h = 0.8$

Water surface is 196.4 + 0.8 = <u>197.2 ft</u> above bottom of aquifer

---

3.26

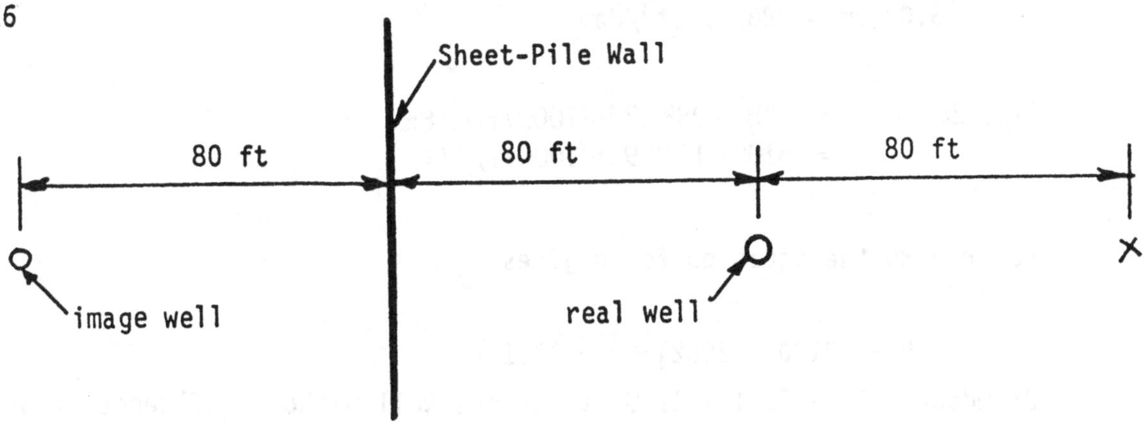

Problem 3-26 Cont'd.

h_0 = 100-10 = 90 ft., K = 0.8 ft/day, r_w = 1 ft.
Q = 7.5 gpm = 1443.8 ft^3/day

Eq. 3-26: $h = [(90)^2 - 1443.8 \ln(1000/r)/0.8\pi]^{1/2}$
$= [8100 - 574.5 \ln(1000/r)]^{1/2}$

a. Real Well	Image Well
r = 80	r = 80
h = 81.5 ft	h = 81.5
	h_0-h = 8.5

Actual GW surface is 81.5 - 8.5 = 73.0 ft above bottom of aquifer

b. Real Well	Image Well
r = 80	r = 240
h = 81.5 ft	h = 85.3
	h_0-h = 4.7

Actual GW surface is 81.5 - 4.7 = 76.8 ft above bottom of aquifer

---

3-27. See Fig. for solution of prob. 3-26.

h_0 = 100 - 10 = 90 ft., K = 0.8 ft/day, r_w = 1 ft.
Q = 15.0 gpm = 2887.7 ft^3/day

Eq. 3-26: $h = [(90)^2 - 2887.7 \ln(1000/r)/0.8\pi$
$= [8100 - 1148.9 \ln(1000/r)]^{1/2}$

For r = 80 the equation for h gives.

$$h = [8100 - 2902]^{1/2} = 72.1 \text{ ft}$$

Drawdown = 90 - 72.1 = 17.9 ft for one well without influence of other
Drawdown = 17.9 + 17.9 = 35.8 ft for two wells.

3-28.

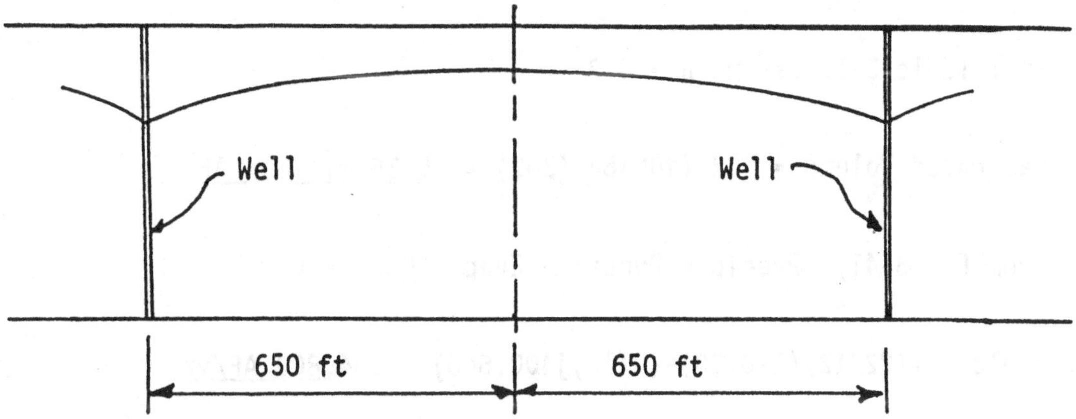

Maximum drawdown will occur at a well.

r_w = 1 ft., Q = 20.0 gpm = 3850 ft^3/day, K = 0.03 ft/day

Drawdown at one well for r = r_w (without others influence)

$h_w = 1200 - 3{,}850/\ln(1000/1)/[2\pi(0.03)800] = 1{,}023.6$

Drawdown = 1200 - 1023.6 = 174.4 ft.

Drawdown at 1300 feet for one well (without other influence) is zero since we have assumed drawdown = 0 for r>1000.

Thus, maximum drawdownn = 176.4 ft and is the same at each well

---

3-29. r_w = 4 ft., Q = 5000 gpm = 962,566 ft^3/day, K = 20 ft/day

ho = 500-100 = 400 ft.

Eq. 3-26: $h = [(400)^2 - 962{,}566 \ln(1000/r)/20\pi]^{1/2}$

$= [160{,}000 - 15{,}320 \ln(1000/r)]^{1/2}$

for each well: r = 150/cos 30° = 173.2 ft.

h = 364.9 ft.

$h_o - h$ = 400-364.9 = 35.1 ft.

Actual drawdown = (35.1 + 35.1 + 35.1) = <u>105.3 ft</u>

3-30. DA = 400 sq. mi., B = 300 ft.

From Table 3-1, assume n = 0.2

Saturated Volume = 0.2 (400)640(300) = <u>15.36 million AF</u>

From Eq. 3-41) Precip - Runoff - Evap - $Qp\Delta t$ = 0

Yield = [(22/12)(1-0.20)-(8/12)]400(640) = <u>204,800 AF/yr</u>

3-31. The piezometric head at the river will be 8.0 ft. and that at the pond will be 3.5. The following grid shows piezometric head h along the boundaries. This grid is used to solve for the interior values of h using Eq. 3-58, but with h substituted for the streamfunction.

(1) h(i,j)=[h(i+1,j)+h(i-1,j)+h(i,j+1)+h(i,j-1)]/4

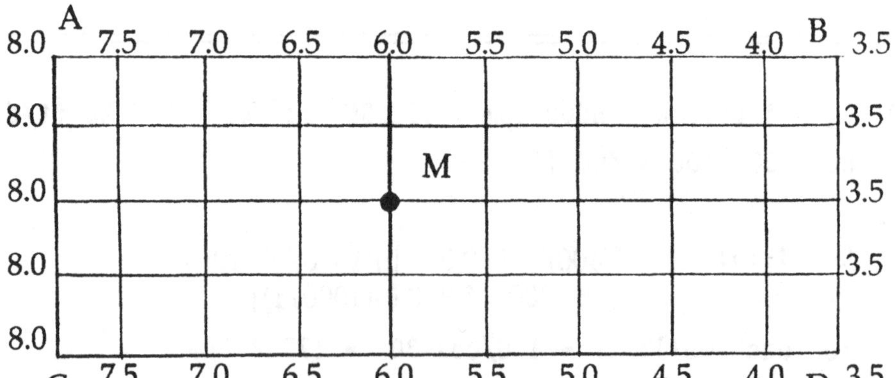

Problem 3-31 Cont'd.

Estimate interior values as equal to boundary values

8.0	7.5	7.0	6.5	6.0	5.5	5.0	4.5	4.0	3.5
8.0	7.5	7.0	6.5	6.0	5.5	5.0	4.5	4.0	3.5
8.0	7.5	7.0	6.5	6.0	5.5	5.0	4.5	4.0	3.5
8.0	7.5	7.0	6.5	6.0	5.5	5.0	4.5	4.0	3,5
8.0									
	7.5	7.0	6.5	6.0	5.5	5.0	4.5	4.0	3.5

Apply Eq. (1) to all interior points one time starting at lower left and them progressing to right and up.

8.0	7.5	7.0	6.5	6.0	5.5	5.0	4.5	4.0	3.5
8.0	7.5	7.0	6.5	6.0	5.5	5.0	4.5	4.0	3.5
8.0	7.5	7.0	6.5	6.0	5.5	5.0	4.5	4.0	3.5
8.0	7.5	7.0	6.5	6.0	5.5	5.0	4.5	4.0	3,5
8.0									
	7.5	7.0	6.5	6.0	5.5	5.0	4.5	4.0	3.5

We obtain no change in interior values. Our initial guess was correct. If we had chosen different initial values in the interior we would have had to apply Eq. 3-58 successively until we finally obtained the values shown.

Problem 3-31 Cont'd.

b) Velocity at M (Darcy Velocity)

Eq. 3-43: $$u = -K\frac{\partial h}{\partial x} = -K\frac{\Delta h}{\Delta x}$$

$$u = -4\,[(5.5 - 6.5)/200)] = \underline{0.02\ \text{ft/day}}$$

c) Determine flow rate

$$Q = uA = 0.02(400)50$$
$$= \underline{400\ \text{ft}^3/\text{day}}$$

3-32. The field is symetrical about AB. Therefore, we can use the field ABCD.

ABD is a streamline on which we can let the streamfunction be zero. EC is another streamline on which the streamfunction will also be constant.

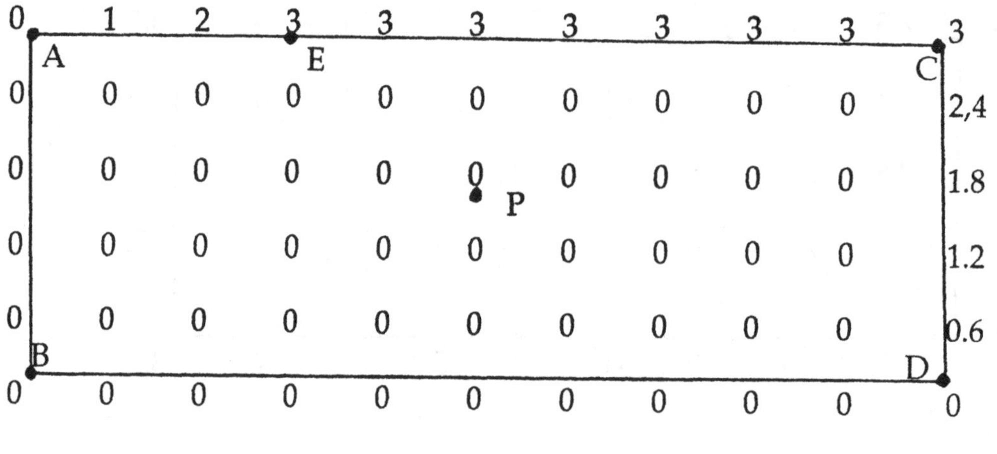

We can arbitrarily set the streamfunction at zero along ABD. The discharge through E is q = V A = (12/12)/60 = 60 ft3/day. We can also arbitrarily set the stream function at 3 along EC since there are three equal increments in ψ between A and E. $\Delta\psi = 3/3 = 1.0$ along AE.

Problem 3-32 Cont'd.

The velocity at point P is calculated from the x and y components of velocity using Eqs. 3-45 and 3-46 respectively.

However, we need to scale the values shown in the figure to relate them to the actual value of q.

The scaling factor is: $q/\psi = 60/3 = 20$

The velocity components are

From Eq. 3-45: $u = 20\ \Delta\psi/\Delta y = 20[1.64 - 1.05 + (1.71 - 1.64)]/2(20) = 0.60$ ft/day

From Eq. 3-46: $v = -20\ \Delta\psi/\Delta x = 20[1.11 - 1.05 + (1.71 - 1.64)]/2(20) = -0.06$ ft/day

The total velocity is

$$V = (u^2 + v^2)^{1/2} = [(0.6)^2 + (0.06)^2]^{1/2}$$
$$= \underline{0.60\ \text{ft/day}}$$

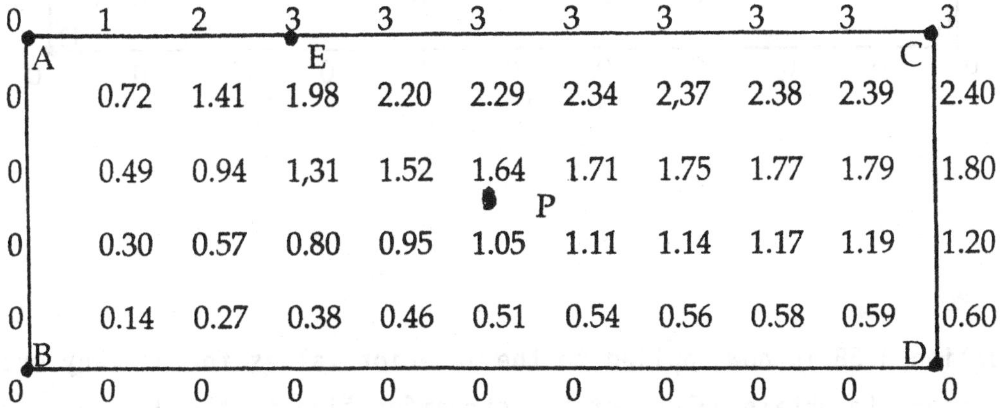

Problem 3-32 Cont'd.

Thus, the stream function between A and E increases uniformly from 0 to 3. Similarly, along DC it also increases uniformly from 0 to 3, but there are five equal increments there so $\Delta\psi = 3/5 = 0.6$ along DC. The previous figure shows the boundary values of the streamfunction for the flow field.

Now estimate interior values of the streamfunction. This can also be done arbitrarily. We will assume that all interior values are zero as shown on the preceding figure. This is not an efficient choice, but if a computer is used there is no real need to refine the choice.

We now use Eq. 3-58 starting from the bottom left interior point and progressing to the right and upward. The following figure shows the result after one complete iteration.

0	1	2	3	3	3	3	3	3	3	3
A			E							C
0	0.25	0.50	0.75	0.75	0.75	0.75	0.75	0.75	1.35	2.4
0	0	0	0	0	0	0	0	0	0.45	1.8
					P					
0	0	0	0	0	0	0	0	0	0.30	1.2
0	0	0	0	0	0	0	0	0	0.15	0.6
B										D
0	0	0	0	0	0	0	0	0	0	0

Equation 3-58 is now applied to the interior values for as many cycles as necessary to obtain values of the streamfunction to the desired numerical accuracy. The following figure shows the interior values of the streamfunction obtained after 17 cycles. The values shown changed less than 0.01 from those of the 16th cycle.

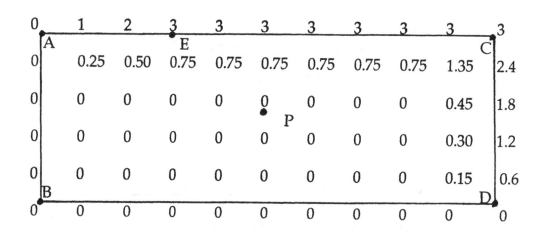

CHAPTER FOUR

4-1. $R = A/P = 4 \times 12/(12 + 2 \times 4) = 2.4$; assume $k_s = 0.003$

$k_s/(4R) = 0.003/(4 \times 2.4) = 0.00031$

$Re\ f^{1/2} = ((4R)^{3/2}/\nu)(2gS)^{1/2} = ((4 \times 2.4)^{3/2}/(1.22 \times 10^{-5}))$

$\times\ (2g \times 10/4{,}000)^{1/2} = 9.8 \times 10^5$; $f = 0.015$ (Fig. 5-4)

$V = \sqrt{8gRS/f} = \sqrt{8g(2.4)10/(0.015(4{,}000))} = 10.15$; $Q = 10.15(4)12 =$ <u>487 cfs</u>

Alternate solution:

$Q = (1.49/n)AR^{2/3}S^{1/2}$ Assume $n = 0.015$

$= (1.49/0.015)\ 4 \times 12(2.4)^{2/3}(10/4{,}000)^{1/2} =$ <u>427 cfs</u>

4-2. $Q = (1.49/n)AR^{2/3}S^{1/2}$

$R = A/P = (\pi D^2/8)/(\pi D/2) = D/4$; $A = \pi D^2/8$

Assume $n = 0.013$

Then $Q = (1.49/0.013) \times (\pi D^2/8)(D/4)^{2/3} \times (1.0/1{,}000)^{1/2}$

$Q =$ <u>22.8 ft³/s</u>

Alternate solution:

$Q = CA\sqrt{RS}$ where $C = \sqrt{8g/f}$; Assume $k_s = 10^{-3}$ ft

$k_s/4R = 10^{-3}/(4 \times D/4) = 2.5 \times 10^{-4}$; Assume $f = 0.016$

Then $C = \sqrt{8 \times 32.2/0.016} = 127$

$Q = 127 \times (\pi \times 4^2/8)\sqrt{1 \times 1/0/1{,}000} = 25.2$ ft³/s; $V = 4.01$ ft/s

$Re = V \times 4R/\nu = 4.01 \times 4 \times 1/(1.2 \times 10^{-5}) = 1.4 \times 10^6$; $f = 0.015$

Solve again: $C = \sqrt{8g/0.015} = 131$; <u>$Q = 26.0$ ft³/s</u>

4-3. $Q = (1/n)AR^{2/3}S^{1/2}$; assume $n = 0.015$

$25 = (1.0/0.015)4d(4d/(4 + 2d))^{2/3} \times 0.002^{1/2}$

Solve by trial and error: <u>d = 2.1 m</u>

4-4. $Q = (1.49/n)AR^{2/3}S^{1/2}$; assume $n = 0.012$

$600 = (1.49/0.012)\ 12d\ (12d/(12 + 2d))^{2/3} \times (10/8{,}000)^{1/2}$

Solve by trial-and-error: <u>d = 5.6 ft</u>

4-5. $R = A/P = (10 + 12)6/(10 + 6\sqrt{5} \times 2) = 132/36.8 = 3.58$

Assume $k_s = 0.003$; $(k_s/4R) = 0.003/(4 \times 3.58) = 0.00021$

$Re\, f^{1/2} = ((4R)^{3/2}/\nu)(2gS)^{1/2} = [(4 \times 3.58)^{3/2}(2g/2{,}000)^{1/2}/(1.41 \times 10^{-5})]$

$= 6.9 \times 10$ Thus, $f = 0.014$

$V = \sqrt{8gRS/f} = \sqrt{8g \times 3.58/(0.014(2{,}000))} = 5.74$ fps

$Q = VA = 5.74(132) =$ 758 cfs

Alternate method, assuming $n = 0.015$

$V = (1.49/n)R^{2/3}S^{1/2} = (1.49/0.015)(3.58)^{2/3}(1/2{,}000)^{1/2} =$ 5.18 fps

$Q = 5.18(132) =$ 684 cfs

4-6. $R = A/P$

where $A = 4 \times 6 + 4 \times 8 = 56\ ft^2$

$P = 6 + 2\sqrt{4^2+8^2} = 23.89$ ft

so $R = 56/23.89 = 2.34$ ft

$Q = (1.49/n)\, AR^{2/3}\, S_o^{1/2}$ (Eq. 4-7a)

Assume $n = 0.012$

Then $S_o^{1/2} = 500 \times (0.012/1.49)/(56 \times 1.763)$

or **$S = 0.00166$**

4-7. Assume $n = 0.012$

$AR^{2/3}/b^{8/3} = Qn/(1.49\, S_o^{1/2}\, b^{8/3})$

$= (3000 \times 0.012)/(1.49 \times (0.001)^{1/2} \times 10^{8/3}$

$= 1.65$

Then from Fig. 4-7, $y/b = 0.90$; $y =$ 9 ft

4-8. Use Fig. 4-7 to solve this problem.

First evaluate $AR^{2/3}/b^{8/3}$

$$AR^{2/3}/b^{8/3} = Qn/(1.49\,S_0^{1/2}\,b^{8/3})$$

$$= 160 \times 0.017/(1.49 \times 0.002^{1/2} \times 8^{8/3})$$

$$= 0.16$$

Entering Fig. 4-7 with a value of 0.16 for $AR^{2/3}/b^{8/3}$ and going vertically upward to the curve of Z = 1.0 we see that we have a value of y/b of approximately 0.35. Therefore the depth is y = 0.35 x 8 or **y = 2.8 ft.** A more precise iterative solution of Eq. (4-8) yields a value of **y = 2.6 ft.**

4-9. Assume k_s = 30cm; R = A/P ≈ 2.21m; $k_s/4R$ = 0.034; f ≈ 0.060 (from Fig. 5-4)

$$C = \sqrt{8g/f} = 36.2\ m^{1/2}s^{-1}$$

$$Q = CA\sqrt{RS} = 332\ m^3/s$$

4-10. Use Fig. 4-7 to solve this problem. Entering Fig. 4-7 with a value of $y/d_0 = 0.8$ we read a value of $AR^{2/3}/d_0^{8/3}$ of approximately 0.3 for the circular conduit. Thus

$$AR^{2/3}/d^{8/3} = Qn/(1.49\,S_0^{1/2}\,d^{8/3})$$

$$0.3 = Qn/1.49\,S_0^{1/2}\,d^{8/3}$$

or $$Q = 0.3 \times 1.49 \times S_0^{1/2} \times d^{8/3}/n$$

Assume n = 0.012

$$Q = 0.3 \times 1.49 \times 0.003^{1/2} \times 5^{8/3}/0.012$$

Q = 149 cfs

A solution of Eq. (4-7a)(by computing A and R for y/d_0 = 0.80) yields

Q = 151 cfs

4.11. Assume n = 0.012

$$AR^{2/3}/b^{8/3} = Qn/(1.49\,S_o^{1/2}b^{8/3})$$

$$= 1{,}000 \times 0.012/(1.49(1/500)^{1/2} \times 10^{8/3})$$

$$= 0.388$$

From Fig. 4-7 with $AR^{2/3}/b^{8/3}$ = 0.388 and z = 1 one finds that y/b = 0.54. Thus y = 0.54 x 10 = **5.40 ft**

4-12. $Q = (1.49/n)AR^{2/3}S^{1/2}$

First calculate A and R from the given figure.

By approximating the area as several triangles and rectangles the area is found to be $A \approx 300\ ft^2$

Likewise, by approximation it is found that $P \approx 125$ ft

Thus, $R = A/P = 2.40$ ft. Assume the same value of n as given in Fig. 4-3 (admittedly this is a fairly gross approximation).

Then, $Q = (1.49/0.038) \times 300 \times 2.4^{2/3} \times .0032^{1/2}$

$= \underline{1193\ cfs}$

==

4-13. $Q = (1.49/n)\ AR^{2/3}\ S^{\ 1/2}$; assume $n = 0.012$

$A = 10 \times 5 + 5^2$, $P = 10 + 2\sqrt{5^2 + 5^2} = 24.14$.ft

$R = A/P = 75/24.14 = 3.107$ ft

Then $Q = (1.49/0.012)\ (75)\ (3.107)^{2/3}\ (5/5280)^{1/2} = \underline{610\ cfs}$

==

4-14. $Q = 100$ cfs; $S = 0.001$; $n = 0.015$

$Q = (1.49/n)\ AR^{0.667}\ S_o^{0.5}$

or $Qn/(1.49S^{0.5}) = AR^{0.667}$

$Qn/(1.49^{0.5}b^{8/3}) = AR^{0.667}/b^{8/3}$

$31.84/b^{8/3} = AR^{0.667}/b^{8/3}$

For different values of b one can evaluate the value of $AR^{0.667}/b^{8/3}$ which one uses to enter Fig. 4-7 from which the values of y/b are found. These values of y/b can then be used to calculate y and A with which the desired graph can be constructed. The pertinent results are shown in the table and figure (below).

4-14. (Continued)

b (ft)	$b^{8/3}$	$\frac{AR^{2/3}}{b^{8/3}}$	y/b	y (ft)	A (ft^2)
2	6.4	5.01	25	50	100
4	40.3	0.79	1.5	6	24
6	119	0.268	0.61	3.7	22
8	256	0.124	0.35	2.8	22.4
10	464	0.069	0.23	2.3	23.3
15	1368	0.023	0.11	1.7	24.8

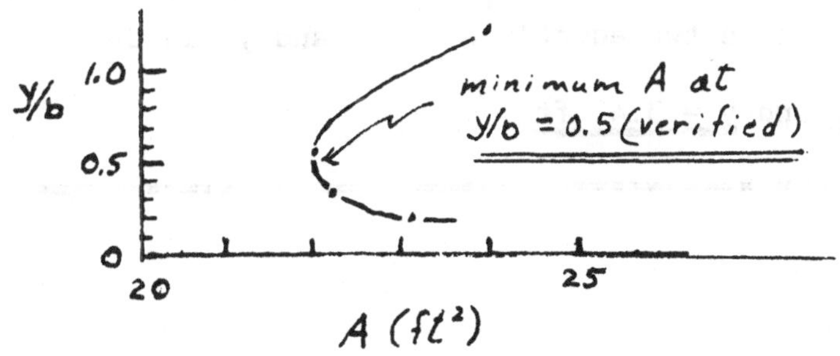

4-15. First compute the mean velocity in the canal and then compare that with the permissible velocities given in Table 4-3. Assume the 1 inch gravel can be considered as fine gravel (n = 0.020).

$V = (1.49/n)\ R^{2/3}\ S^{1/2}$

where R = A/P; $A = 60 + 6 \times 12 = 132\ ft^2$

$P = 10 + 2\ \sqrt{6^2 + 12^2} = 36.8$ ft; so $R = 132/36.8 = 3.58$ ft

Then $V = (1.49/0.020)\ (3.58)^{0.667}\ (0.004)^{0.5} = 11.0$ ft/sec

The permissible velocity for fine gravel is 2.50 ft/sec (Table 4-3); therefore, the gravel can be expected to be scoured by the flow.

4-16. From Table 4-3 the maximum permissible velocity for coarse gravel is given as 4.00 ft/sec and n = 0.025.

Manning equation: $V = (1.49/n)\ R^{2/3}\ S^{1/2}$

or $R^{2/3} = V/((1.49/n)\ (S)^{1/2}) = 4.0/(1.49/0.025)\ (0.016)^{1/2}$

$= 1.678$; $R = 2.173$ ft

Also $A = Q/V = 200/4.0 = 50\ ft^2$

Assume side slopes will be <u>1 vert. to 2 horiz.</u>

$P = A/R = 50/2.173 = 23.01$ ft

Also: $P = b\ \ 2\sqrt{y^2 + 4y^2} = 23.01$ ft

$A = by + 2y^2 = 50\ ft^2$

Solving the above two equations for b and y yields

<u>b = 7.59 ft</u> and <u>y = 3.45 ft</u>

4-17. From Table 4-3: V = 1.75 ft/sec; n = 0.020

Following the procedure of problem 4-16 we have:

$R^{2/3} = V/((1.49/n)\ (S)^{1/2}) = 1.75/((1.49/0.020)\ (0.0005)^{0.5})$

$= 1.051$; $R = 1.077$ ft

$A = Q/V = 30/1.75 = 17.14\ ft^2$

Assume side slopes are to be 1 V to 3 H (Table 4-2)

$P = A/R = 17.14/1.077 = 15.9$ ft

$P = b + 2\ \sqrt{y^2 + (3y)^2} = 15.9$ ft

$A = by + 3y^2 = 17.14\ ft^2$

Solving the above two equations yields

<u>y = 1.65 ft</u> and <u>b = 5.43 ft</u>

4-18. $V = Q/A = 500/(10 \times 3) = 16.67$ ft/s

$F = V/\sqrt{gy} = 16.67/\sqrt{32.2 \times 3} = \underline{1.7 \text{ (supercritical}}$

4-19. $V = Q/A;\ A = Q/V = 10/1 = 10\ m^2;\quad d = 10/5 = 2m$

$F = V/\sqrt{gy} = 1/\sqrt{9.81 \times 2} = \underline{0.22 \text{ (subcritical)}}$

4-20. $F = V/\sqrt{gy} = (Q/A)/\sqrt{gy} = (Q/(By))/\sqrt{gy}$

$F = Q/(B\sqrt{g}\ y^{3/2})$

$F_{0.3} = 12/(3\sqrt{9.81} \times 0.3^{3/2}) = \underline{7.77 \text{ (supercritical}}$

$F_{1.0} = 12/(3\sqrt{9.81} \times 1.0^{3/2}) = \underline{1.28 \text{ (supercritical)}}$

$F_{2.0} = 12/(3\sqrt{9.81} \times 2.0^{3/2}) = \underline{0.452 \text{ (subcritical)}}$

$y_c = (q^2/g)^{1/3} = ((Q/B)^2/9.81)^{1/3} = \underline{1.18\ m}$

4-21. $E_{0.3} = y_1 + q^2/(2gy_1^2) = 0.30 + (12/3)^2/(2 \times 9.81 \times 0.3^2) = \underline{9.36\ m}$

Then $y_2 + q^2/(2gy_2^2) = 9.36$ m where $q = (12/3)\ m^2/s$

Solving: $y_{alt.} = \underline{9.35\ m}$

4-22. Check Froude number:

$Fr = V/\sqrt{gy} = 6/\sqrt{0.1 \times 9.81} = 6.06$

The Froude number is greater than 1 so the flow is supercritical.

$E = y + V^2/2g$

$E = 0.1 + 6^2/(2 \times 9.81) = 1.935$ m

Solving for the alternate depth for an E of 1.935 yields $y_{alt.} = \underline{1.93\ m}$

4.23. $V_c^2/g = y_c;\quad y_c = \underline{1.12\ ft}$

4-24. $Q = (1/n)AR^{2/3}S^{1/2}$

$9 = (1/0.014) \times 4y(4y/(B+2y))^{2/3} \times (0.005)^{1/2}$

Solving for y gives: $y = 0.693$ m and $V = Q/(By) = 3.25$ m/s

Then $F = V/\sqrt{gy} = \underline{1.24 \text{ (supercritical)}}$

4-25.

y (m)	E (m)
0.25	7.59
0.30	5.40
0.40	3.27
0.50	2.33
0.60	1.87
0.70	1.64
0.80	1.52
0.90	1.47
1.00	1.46
1.10	1.48
1.40	1.63
2.00	2.11
4.00	4.03
7.00	7.01

$E = y + q^2/(2gy^2)$ (for a rectangular channel)

For this problem $Q = Q/B = 18/6 = 3\text{m}^2/\text{s}$

So $E = y + 3^2/(2gy^2)$

or $E = y + 0.4587/y^2$

E vs. y is shown in table at left.

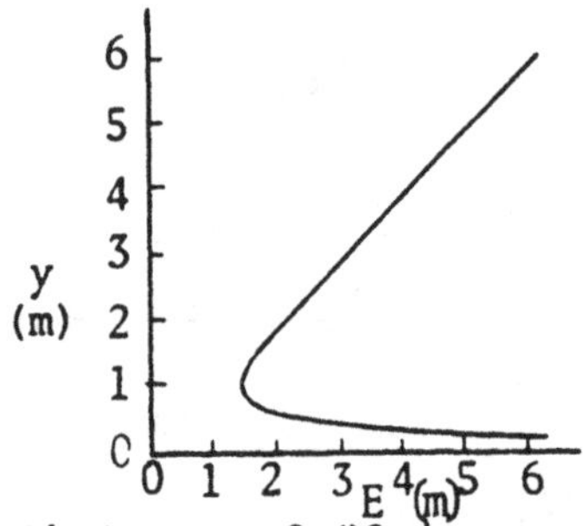

The alternate depth to y = 0.30 is $\underline{y = 5.38 \text{ m}}$

Sequent depth: $y_2 = (y_1/2)(\sqrt{1 + 8F_1^2} - 1)$

$F_1 = V/\sqrt{gy_1} = (3/0.3)/\sqrt{9.81 \times 0.30} = 5.83$

Then $y_2 = (0.3/2)(\sqrt{1 + 8 \times 5.83^2} - 1) = \underline{2.33 \text{ m}}$

4-26. $d_{brink} \approx 0.71\, y_c = 0.71(q^2/g)^{1/3}$

Then for $d_{brink} = 0.25$ m $\quad q = (0.25 \times g^{1/3}/0.71)^{3/2}$

$q = 0.654 \text{ m}^2/\text{s}$

Then $\underline{Q = 3q = 1.96 \text{ m}^3/\text{s}}$

4-27.

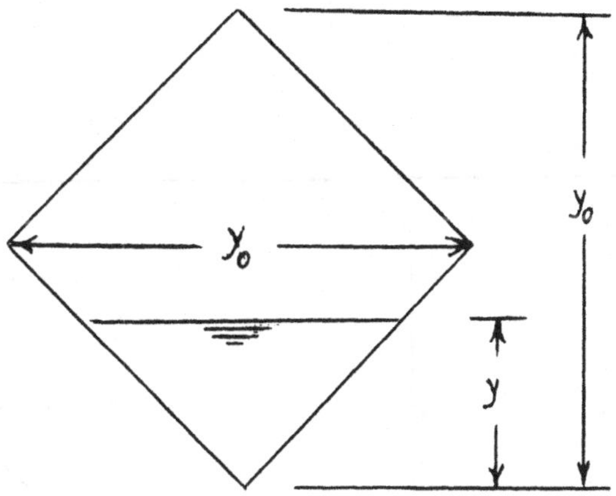

Up to half depth the cross-sectional-flow area will be

$$A = y^2 = ky_0^2 \tag{1}$$

where $k = y/y_0$

From half depth to full flow the cross-sectional-flow area will be

$$A = 0.25y_0^2 + (y - 0.5y_0)y_0 - (y - 0.5y_0)^2$$

$$A = (-k^2 + 2.0k - 0.5)y_0^2 \tag{2}$$

Also $((1/2)P)^2 = 2y^2$

or $P = 2\sqrt{2}\,y = 2\sqrt{2}ky$

Then $R = A/P = (-k^2 + 2k - 0.5)y_0^2/(2\sqrt{2}k)$ (3)

Then using Eqs. (1), (2) and (3) we can compute $AR^{2/3}/A_0R_0^{2/3}$ and $R^{2/3}/R_0^{2/3}$. The tabular and graphical results are shown below.

$k = y/y_0$	R/R_0	$(R/R_0)^{2/3}$	A/A_0	$(A/A_0)(R/R_0)^{2/3}$
0.00	0.00	0.00	0.00	0.00
0.10	0.20	0.34	0.02	0.01
0.20	0.40	0.54	0.08	0.04
0.30	0.60	0.71	0.19	0.13
0.40	0.80	0.86	0.32	0.27
0.50	1.00	1.00	0.50	0.50
0.60	1.13	1.09	0.68	0.73
0.70	1.17	1.11	0.82	0.91
0.80	1.15	1.10	0.92	1.01
0.90	1.09	1.06	0.98	1.04
1.00	1.00	1.00	1.00	1.00

4-27. (Continued)

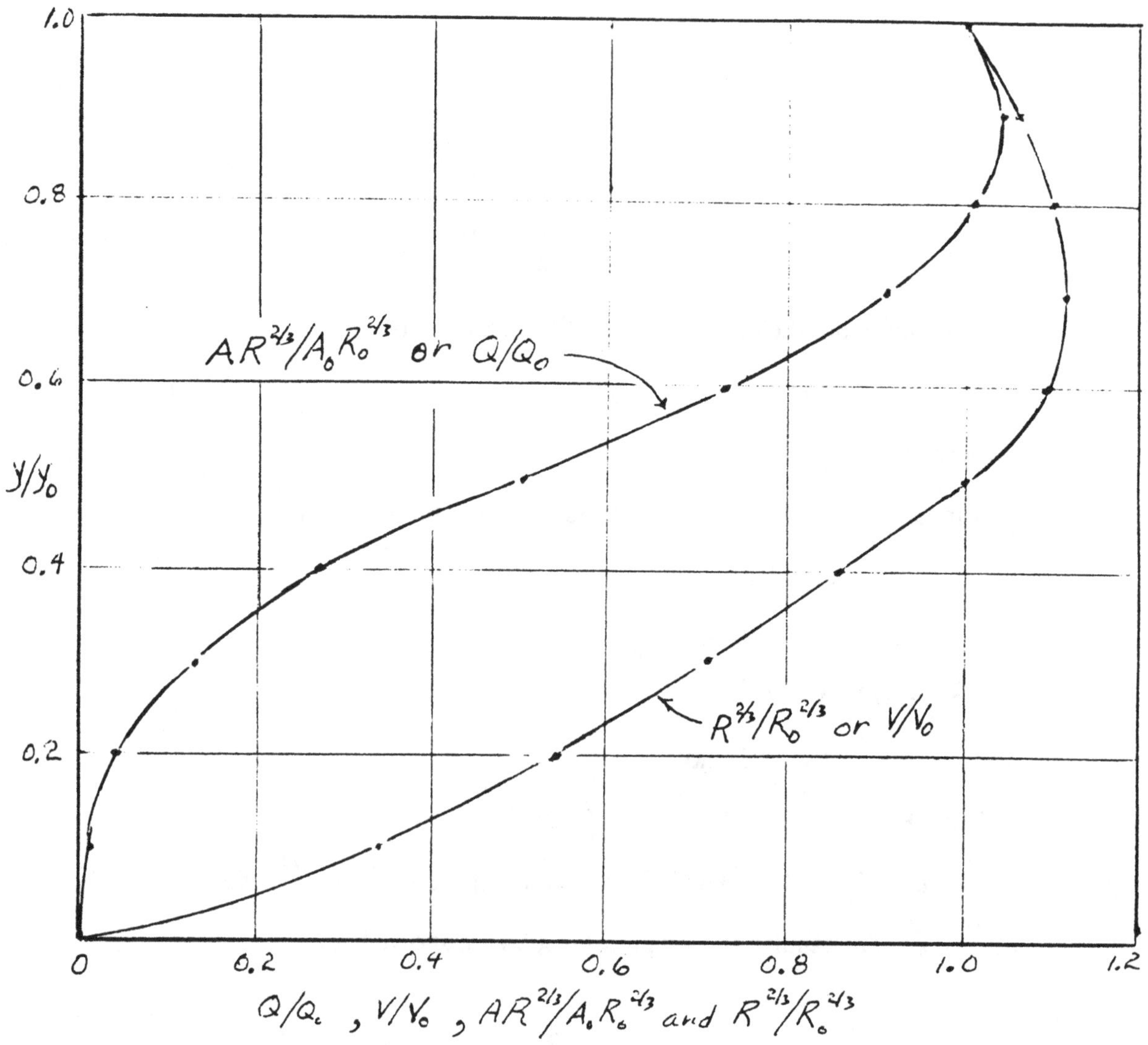

4-28. Solution like that for P4-26:

$q = ((1.20 \times (32.2)^{1/3}/0.71)^{3/2}$

$q = 12.47\ m^2/s$

Then $Q = 15 \times 12.47 =$ 187 cfs

4-29. For critical flow to prevail this equality must exist: $A^3/T = Q^2/g$

For this problem Q = 700 ft^3/sec, so $A_c^3/T_c = (700)^2/32.2 = 15{,}217$

Because $A = 10y + 2y^2$ and $T = 10 + 4y$ one can

solve for y that will satisfy the equation. the solution

yields $y_c = 4.06$ ft for which A = 73.6 ft^2 and T = 26.3 ft

---

4-30. $A_c^3/T = Q^2/g$

$= 25^2/32.2 = 19.41$

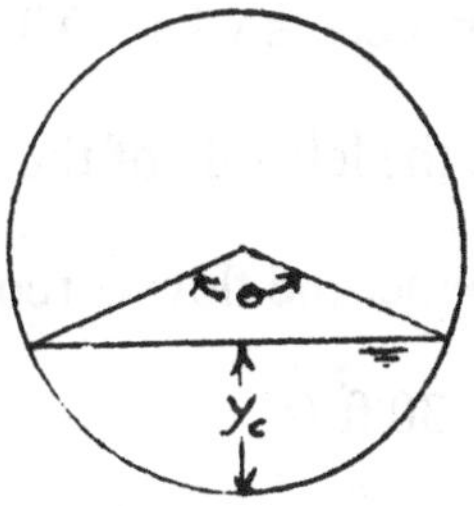

It can be shown that

$A = (D^2/8)(\theta - \sin\theta)$

and $T = D \sin(\theta/2)$

Then $A_c^3/T_c = D^5 \ (\theta - \sin\theta)^3 / (8^3 \times \sin(\theta/2)) = 19.41$

Solving the above equation for θ yields θ = 150°

from which it can be found (by trigonometry)

that $y_c = 0.371D = 1.484$ ft

4-31. $A_c^3/T_c = Q^2/g$

where for this channel $A_c = d_c^2$ and $T_c = 2d_c$

thus $(d_c^2)^3/2d_c = Q^2/g$

or $d_c = (2Q^2/g)^{0.2}$

4-32. $E = y + V^2/(2g)$

$y = \text{depth} = (3/2)\tan 45^{\circ} = 1.5$ ft

$V = Q/A = 30/(1/2 \times 3 \times 1.5)$

$= 30/2.25$

$= 13.33$ ft/s

Then $E = 1.5 + 13.33^2/64.4$

$=$ **<u>4.26 ft</u>**

Critical flow occurs when $Q^2T/(gA^3) = 1$

so evaluate the left side of the equation to see if it is greater or less than unity to discern whether the flow is respectively supercritical or subcritical.

$Q = 30$ ft^3/s

$T = 3$ ft

$g = 32.2$ ft/s^2

$A = 2.25$ ft^2 (from calculation for E, above)

Therefore $Q^2T/(gA^3) = 30^2 \times 3/(32.2 \times 2.25)$

$= 37.27$ (much greater than unity)

<u>The flow is supercritical</u>

4-33. Write the energy equation from the water surface in the reservoir to the water surface in the canal where the depth = 4 ft. Let datum be at the bottom of the canal.

$V_o^2/2g + y_c = V_1^2/2g + y$

V_o = velocity in reservoir = 0

$y_c = 4 + 0.1 = 4.1$ ft

$y_1 = 4$ ft

Then $4.1 = V_1^2/2g + 4.0$

$V_1 = (0.1 \times 2g)^{1/2}$

$= 2.538$ ft/s

$q = Vy_1$

$= 2.538 \times 4$

$= \underline{10.15\ ft^2/s}$

Upstream of the break in grade the cross-sectional area of flow depends in part on the velocity at the section; therefore, the energy and continuity equations must be solved to obtain the water surface elevation and velocity at each section. However, there is very little water surface elevation change between the reservoir and the canal. Therefore, as a very good approximation for the velocity calculation, one can assume a straight line change in water surface elevation between reservoir and canal. Thus, the water surface elevations and depths at sections 2 ft, 4 ft and 8 ft upstream of the break in grade are given below. Assume negligible velocity at 12 ft from break.

Section[1]	W.S. elev above Canal W.S.[2]	Flow Depth[3]	V	$V^2/2g$	W.S. elev. above W.S. elev in canal[4]
2	0.017 ft	4.217 ft	2.413 ft/s	0.0904 ft	0.0096 ft
4	0.033	4.433	2.29	0.0814	0.0186
8	0.067	4.867	2.085	0.068	0.032
Reservoir	0.1	—	0	0	0.1

1. Feet from break in grade
2. Based upon assumed linear change in water surface elevation between reservoir and canal.
3. Flow depth is based upon assumed W.S. elevations in Column 2.
4. Obtained by writing the energy equation between the section and canal using the velocity given in column 4. Further iterations produce negligible difference in results (less than 0.01 ft in water surface elevations).

4-34. Use the momentum equation written from a section upstream of the angle irons to a section downstream of them. Write it per foot of width of channel.

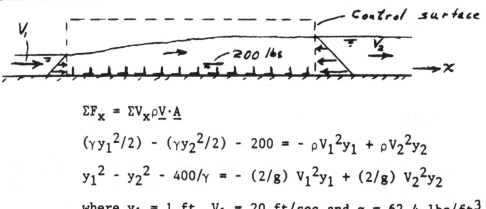

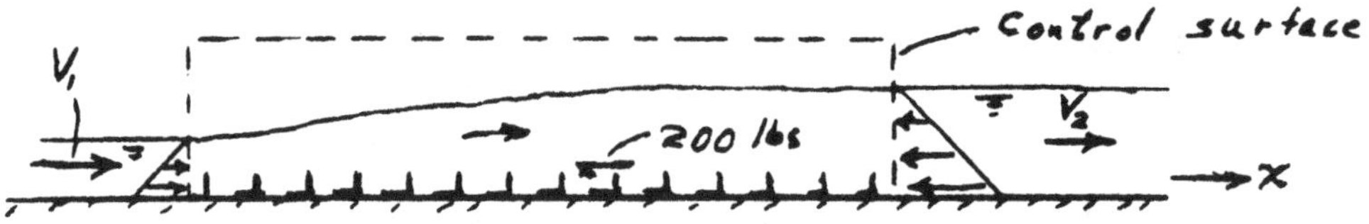

$\Sigma F_x = \Sigma V_x \rho \underline{V} \cdot \underline{A}$

$(\gamma y_1^2/2) - (\gamma y_2^2/2) - 200 = -\rho V_1^2 y_1 + \rho V_2^2 y_2$

$y_1^2 - y_2^2 - 400/\gamma = -(2/g)\, V_1^2 y_1 + (2/g)\, V_2^2 y_2$

where $y_1 = 1$ ft, $V_1 = 20$ ft/sec and $\alpha = 62.4$ lbs/ft^3

The above equation simplifies to

$y_2^2 + (24.84/y_2) - 19.43 = 0$

Solving yields $\underline{y_2 = 1.43 \text{ ft}}$

4-35. $V_1 = 3$ m/s So $E_1 = y_1 + V_1^2/2g = 3 + 3^2/(2 \times 9.81) = 3.46$ m

$F_1 = V_1/\sqrt{gy_1} = 3/\sqrt{9.81 \times 3} = 0.55$ (subcritical)

Then $E_2 = E_1 - \Delta z_{step} = 3.46 - 0.30 = \underline{3.16 \text{ m}}$

$y_2 + q^2/(2gy_2^2) = 3.16$ m

$y_2 + 9^2/(2gy_2^2) = 3/16$

$y_2 + 4.13/y_2^2 = 3.16$

Solving for y_2 yields $y_2 = 2.50$ m

Then $\Delta y = y_2 - y_1 = 3.00 - 2.50 = \underline{-0.50}$

W.S. drops $\underline{0.20 \text{ m}}$

For a downward step $E_2 = E_1 + \Delta z_{step} = 3.46 + 0.3 = \underline{3.76 \text{ m}}$

$y_2 + 4.13/y_2^2 = 3.76$ Solving: $y_2 = 3.40$ m

Then $\Delta y = y_2 - y_1 = 3.40 - 3 = \underline{0.40 \text{ m}}$

W.S. elevation change = $\underline{+0.10 \text{ m}}$

4-35. (continued)

Max. upward step before altering upstream conditions:

$y_c = y_2 = \sqrt[3]{q^2/g} = \sqrt[3]{9^2/9.81} = 2.02$

$E_1 = \Delta z_{step} + E_2$ where $E_2 = 1.5\ y_c = 1.5 \times 2.02 = 3.03$ m

Max. $z_{step} = E_1 - E_2 = 3.46 - 3.03 = \underline{\underline{0.43\ m}}$

4-36.

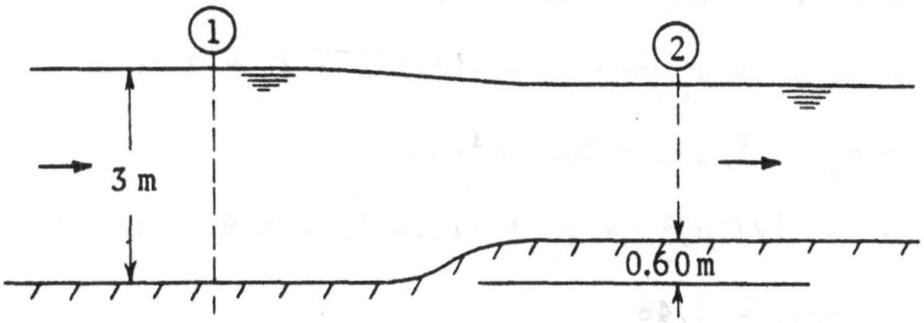

$E_2 = E_1 - 0.60$; $V_1 = 2$ m/s; $F_1 = V_1/\sqrt{gy_1} = 2/\sqrt{9.81 \times 3} = 0.369$

Then $E_2 = (3 + (2^2/(2 \times 9.81)) - 0.60 = 2.60$ m

Solve for y_2: $y_2 + q^2/(2gy_2^2) = 2.60$ where $q = 2 \times 3 = 6 m^3/s/m$

Then $y_2 + 6^2/(2 \times 9.81 \times y_2^2) = 2.60$

$y_2 + 1.83/y_2^2 = 2.60$

Solving, one gets $y_2 = 2.24$ m; $\Delta y = y_2 - y_1 = 2.24 - 3.00 = \underline{\underline{-0.76m}}$

Water surface drops $\underline{\underline{0.16\ m}}$

For downward step of 15 cm we have

$E_2 = (3 + (2^2/(2 \times 9.81)) + 0.15 = 3.35$ m

$y_2 + 6^2/(2 \times 9.81 \times y_2^2) = 3.35$

$y_2 + 1.83/y_2^2 = 3.35$

Solving: $y_2 = 3.17$ m or $y_2 - y_1 = 3.17 - 3.00 = \underline{\underline{+0.17\ m}}$

Water surface rises $\underline{\underline{0.02\ m}}$

The maximum upstep possible before affecting upstream water surface levels is for $y_2 = y_c$

$y_c = \sqrt[3]{q^2/g} = 1.54$ m

4-36. (continued)

Then $E_1 = \Delta z_{step} + E_{2,crit}$

$\Delta z_{step} = E_1 - E_{2,crit} = 3.20 - (y_c + V_c^2/2g) = 3.20 = 1.5 \times 1.54$

$z_{step} = \underline{\underline{+0.89\ m}}$

4-37. $F_1 = V_1/\sqrt{gy_1} = 3/\sqrt{9.81 \times 3} = 0.55$ (subcrit)

$E_1 = E_2 = y_1 + V_1^2/2g = 3 + 3^2/\sqrt{2 \times 9.81} = 3.46$ m

$q_2 = Q/B_2 = 27/2.6 = 10.4\ m^3/s/m$

Then $y_2 + q^2/(2gy_2^2) = y_2 + (10.4)^2/(2 \times 9.81 \times y_2^2) = 3.46$

$y_2 + 5.50/y_2^2 = 3.46$

Solving: $y_2 = 2.71$ m

$\Delta z_{water\ surface} = \Delta y = y_2 - y_1 = 2.71 - 3.00 = \underline{\underline{0.29\ m}}$

Max. contraction without altering the upstream depth will occur with $y_2 = y_c$

$E_2 = 1.5\ y_c = 3.46;\quad y_c = 2.31$ m

Then $V_c^2/2g = y_c/2 = 2.31/2$ or $V_c = 4.76$ m/s

$Q_1 = Q_2 = 27 = B_2 y_c V_c;\quad B_2 = 27/(2.31 \times 4.76) = 2.46$ m

The width for max. contraction = $\underline{\underline{2.46\ m}}$

4-38. First determine the discharge in the channel.

$Q = (1.49/n)\ AR^{2/3}S_o^{1/2}$; assume $n = 0.015$

$A = 10 \times 5 + 5 \times 10 = 100\ ft^2$

$P = 10 + 2\ (5^2 + 10^2)^{0.5} = 32.36$ ft

$A/P = 100/32.36 = 3.090$ ft

$Q = (1.49/0.015)\ (100)\ (3.090)^{0.667} \times (0.0005)^{0.5}$

$= 471$ cfs; $V_{channel} = Q/A_c = 4.71$ ft/sec

Next, determine cross section of flume

Given: $b = y$ for flume; $F_r \leq 0.60$

$V_f/\sqrt{gy_f} = 0.60$

$V_f = 0.60\ \sqrt{gy_f}$

Also $Q = 471 = V_fA_f$

$471 = V_f\ y_f^2$

Solve (1) and (2) for y_f; $y_f = 7.178$ ft; $A_f = 51.5\ ft^2$

For design purposes let $y_f = 7.2$ ft and $b_f = 7.1$ ft

Then $V_f = 471\ cfs/51.5\ ft^2 = 9.15$ ft/sec

The profile of the channel, transition and flume will be as shown below

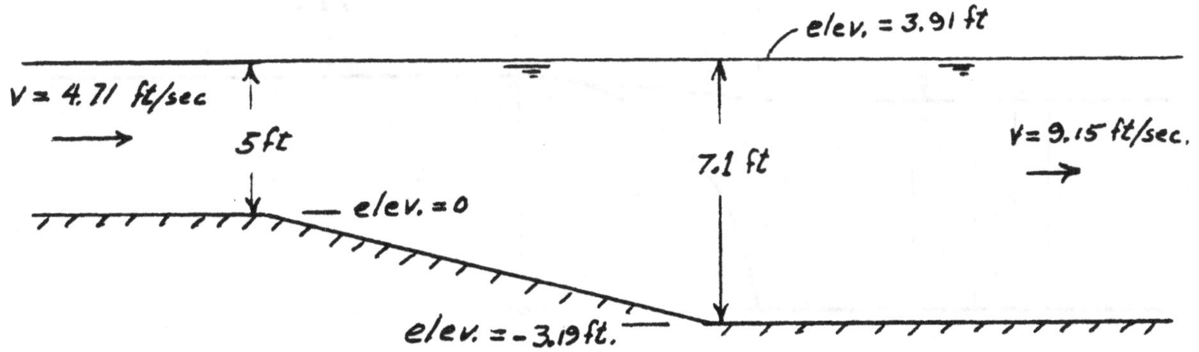

4-38. (continued)

Determine water surface elevation at upstream end of flume:

Write the energy equation from the channel to the flume assuming

$\alpha_1 = \alpha_2 = 1.0$. Assume the elevation of the channel bottom = 0.

$(V_c^2/2g) + z_c = (V_f^2/2g) + z_{W.S.}$ in flume $+ \Sigma h_L$

where $z_c = 5$ ft

$V_c = 4.71$ ft/sec

$V_f = 9.15$ ft/sec

$\Sigma h_L = KV_f^2/2g$ (assume K = 0.10)

Then $z_{W.S., \text{ flume}} = (4.71^2/2g) + 5 - (9.15^2/2g) - 0.1\ (9.15^2/2g)$

$= 3.91$ ft

Then the invert elevation of flume = 3.91 ft - 7.1 f = -3.19 ft

That elevation is shown on the above figure.

The plan view of the transition is shown below

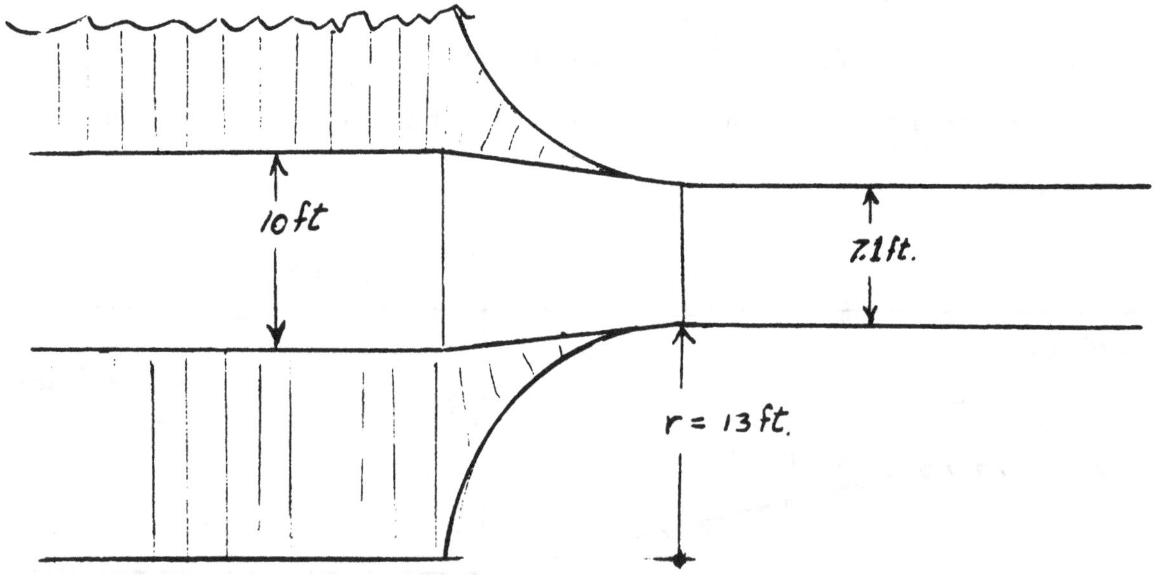

Completion of design will involve decision about freeboard (perhaps 1 ft) and details involved with transition bottom as shown in sketch (above).

4-39. $y_0 + q^2/(2gy_0^2) = y_1 + q^2/(2gy_1^2)$; $q = 1.2\ m^3/s/m$; $y_0 = 5$ m

$5 + 1.2^2/(2(9.81)5^2) = y_1 + 1.2^2/(2(9.81)y_1^2) \rightarrow y_1 = 0.123$ m

$F_1 = q/\sqrt{gy_1^3} = 2/\sqrt{9.81(0.123)^3} = 8.88$

$y_2 = (y_1/2)(\sqrt{1 + 8F_1^2} - 1) = (0.123/2)(\sqrt{1 + 8(8.88^2)} - 1) = \underline{1.48m}$

4-40. $F_1 = q/\sqrt{gy^3} = 1.9/\sqrt{9.81(0.3)^3} = 3.7 > 1$

$\therefore$ Jump can form.

$y_2 = (y_1/2)(\sqrt{1 + 8F_1^2} - 1) = (0.3/2)(\sqrt{1+8(3.7)^2} - 1) = \underline{1.43\ m}$

4-41. First develop the expression for y_1 and $V_{theor.}$ Write the energy equation from the upstream pool level to y_1.

$V_o^2/2g + Z_o = V_1^2/2g + z_1$; assume V_o is negligible

$0 + 100 = V^2_{theor.}/2g + y_1$

But $V_{theor.} = V_{act.}/0.95$

and $V_{act.} = q/y_1$

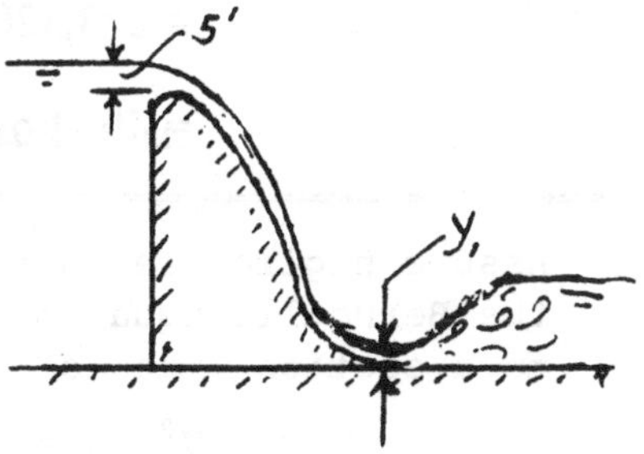

Consider a unit width of spillway. Then

$q = Q/L = K\sqrt{2g}\ H^{1.5}$

$= 0.5\sqrt{2g}\ (5^{1.5})$

$= 44.86$ cfs/ft

Solving Eqs. (1), (2), and (3) yields

$y_2 = 0.59$ ft and $V_{act.} = 76.03$ ft/sec

$Fr_1 = V/\sqrt{gy_1} = 76.03/\sqrt{(32.2)(0.59)} = 17.44$

Now solve for the depth of flow on the apron:

$y_2 = (y_1/2)((1 + 8\ Fr_1^2)^{0.5} - 1)$

$= (0.59/2)(1 + 8(17.44^2))^{0.5} - 1) = \underline{14.3\ ft}$

4-42. $y_2/y_1 = 8 = 1/2(\sqrt{1 + 8\,Fr_1^2} - 1)$

$Fr_1 = 6.00$

But $Fr_1 = V/\sqrt{gy}$

or $V_1 = 6.00\sqrt{32.2 \times 1}$

$= 34.05$ ft/s

and $V_2 = (1/8)V_1 = 4.26$ ft/s

$h_L = E_1 - E_2$

$= (y_1 + V_1^2/2g) - (y_2 + V_2^2/2g)$

$= (1 + 18) - (8 + 0.28)$

$= 10.72$ ft

Power Loss $= Q\gamma h_L$

$= (34.05 \times 12)(62.4)(10.72)$

$= 273{,}320$ ft·lb/s

= 497 horsepower

4-43. Assume negligible energy loss for flow under the sluice gate. Write the Bernoulli equation from a section upstream of the sluice gate to a section immediately downstream of the sluice gate.

$$y_0 + V_0^2/2g = y_1 + V_1^2/2g$$

$$65 + \text{neglig.} = 1 + V_1^2/2g$$

$$V_1 = \sqrt{64 \times 64.4} = 64.2 \text{ ft/s}$$

$$F_1 = V_1/\sqrt{gy_1} = 64.2/\sqrt{32.2 \times 1} = 11.3$$

Now solve for the depth after the jump:

$$y_2 = (y_1/2)(\sqrt{1 + 8F_1^2} - 1)$$

$$= (1/2)(\sqrt{1 + 8 \times 11.3^2} - 1) = 15.5 \text{ ft}$$

$$h_L = (y_2 - y_1)^3/(4y_1y_2)$$

$$= (15.51)^3/(4 \times 1 \times 15.51) = \underline{49.2 \text{ ft}}$$

$$P = Q\gamma h_L/550$$

$$= (64.2 \times 1 \times 6) \times 62.4 \times 49.2/550 = \underline{2{,}150 \text{ horsepower}}$$

4-44 $y_2 = (y_1/2)(\sqrt{1 + 8\,Fr_1^2} - 1)$ (1)

where $y_1 = 1$ ft

$Fr_1 = V_1/\sqrt{gy_1}$

$= 2.82$

Then solving Eq. (1) yields **$y = 3.519$ ft**

$V = (1/3.519)(16) = 4.55$ ft/s

$h_L = E_1 - E_2$

$= (y_1 + V_1^2/2g) - (y_2 + V_2^2/2g)$

$= (1 + 3.98) - (3.52 + 0.32)$

$=$ **1.14 ft**

4-45. $V = (1/n)R^{2/3}S_0^{1/2}$ where $n = 0.015$ (assume)

$R = A/P = (0.4 \times 10)/(2 \times 0.4 + 10) = 0.370$ m

Then $V = (1/0.015)(0.370)^{2/3} \times (0.04)^{1/2} = 6.87$ m/s

Then $F_1 = V/\sqrt{gy_1} = 6.87/\sqrt{9.81 \times 0.40} = 3.47$ (supercritical)

Then $y_2 = (y_1/2)(\sqrt{1+8 \times F_1^2} - 1) = (0.40/2)(\sqrt{1+8 \times (3.47)^2} - 1) = $ 1.77 m

---

4-46. Assume the shear stress will be the averate of τ_{0_1}, uniform approaching the jump, and τ_{0_2}, uniform flow leaving the jump.

$\tau_0 = f\rho V^2/8$

where $f = f(Re, k_s/4R)$

$R_{e_1} = V_1(4R_1)/\nu$

From sol. to P. 4-45:

$R_{e_1} = 6.87 \times (4 \times 0.37)/10^{-6}$

$R_{e_1} = 10^7$

assume $k_s = 3 \times 10^{-3}$m

$k_s/4R_1 = 3 \times 10^{-3}/(4 \times 0.37)$

$k_s/4R_1 = 2 \times 10^{-3}$

From Fig. 5-4, $f_1 = 0.024$

Then

$\tau_{0_1} = 0.024 \times 1{,}000 \times (6.87)^2/8$

$\tau_{0_1} = 142$ N/m^2

$R_{e_2} = V_2 \times (4R_2)/\nu$

$V_2 = V_1 \times 0.4/1.77 = 1.55$ m/s

$R_2 = A/P = (1.77 \times 10)/(2 \times 1.77 + 10) = 1.31$ m

$R_{e_2} = 1.55 \times (4 \times 1.31)/10^{-6}$

$R_{e_2} = 8 \times 10^6$

$k_s/4R_2 = 3 \times 10^{-3}/(4 \times 1.31)$

$k_s/4R_2 = 6 \times 10^{-4}$

$f_2 = 0.018$

$\tau_{0_2} = 0.018 \times 1{,}000 \times (1.55)^2/8$

$\tau_{0_2} = 5.4$ N/m^2

$$\tau_{avg} = (142 + 5.4)/2 = 74 \text{ N/m}^2$$

Then $F_s = \tau_{avg}A_s = \tau_{avg}PL$

where $L \approx y_2$, $P \approx B + (y_1+y_2)$

Then $F_s \approx 74(10+(0.40+1.77))(6 \times 1.77) = 9{,}560$ N

$F_H = (\gamma/2)(y_2^2-y_1^2)B = (9{,}810/2)((1.77)^2 - (0.40)^2) \times 10 = 145{,}820$ N

Thus, $F_s/F_H = 9{,}560/145{,}820 = \underline{0.066}$

Note: The above estimate probably gives an excessive amount of wgt. to τ_{0_1} because τ_0 will not be linearly distributed. A better estimate might be to assume a linear distribution of velocity with an average f and integrate $\tau_0 dA$ from one end to the other.

4-47. $q = 0.40 \times 10 = 4.0$ m^3/s/m

Then $y_c = \sqrt[3]{q^2/g} = \sqrt[3]{(4.0)^2/9.81} = 1.18$ m

Then we have $y < y_n < y_c$; therefore, the water surface profile will be an $\underline{S3}$.

4-48. $q = 5/3 \quad F_1 = q/\sqrt{gy^3} = (5/3)/\sqrt{9.81(0.3)^3} = 3.24 > 1$ (supercritical)

Flow over weir, $Q = (0.40+0.05\ H/P)L\sqrt{2g}\ H^{3/2}$

$$5 = (0.40+0.05\ H/1.6)\times 3\sqrt{2(9.81)}H^{3/2}$$

Solving by iteration gives H = 0.917 m

Depth upstream of weir = 0.917 + 1.6 = 2.52 m

$F_2 = (5/3)/\sqrt{9.81(2.52)^3} = 0.133 < 1$ (subcritical)

∴ A hydraulic jump forms. $y_2 = (0.3/2)(\sqrt{1+8(3.24)^2}-1) = 1.23$ m

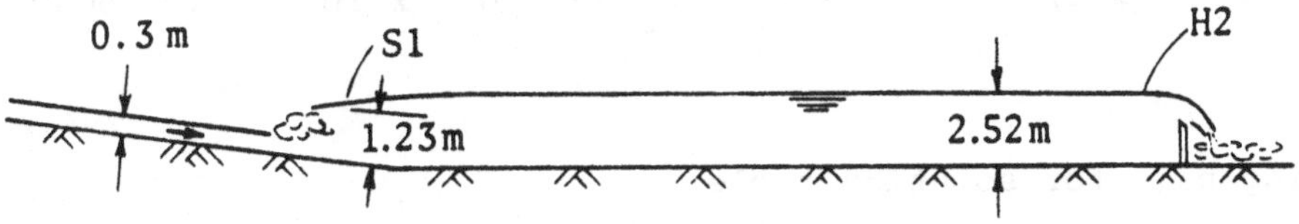

4-49. $F_1 = q/\sqrt{gy^3} = 3/\sqrt{9.81(0.2)^3} = 10.71$

$F_2 = 3/\sqrt{9.81(0.6)^3} = 2.06$

∴ Continuous H-3 profile

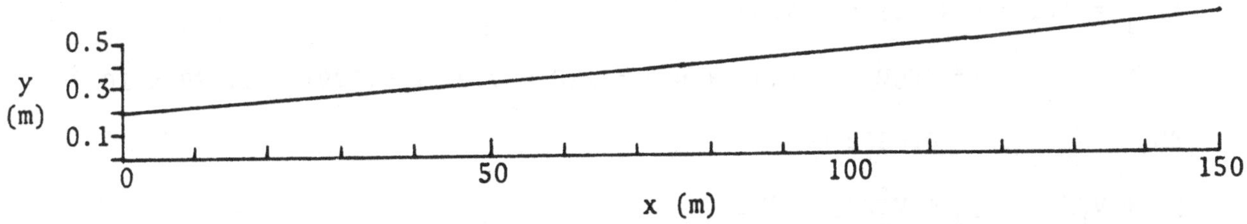

y	$\bar{y}$	V	$\bar{V}$	E	ΔE	S_f	Δx	x
0.2		15		11.6678				0
	0.25		12.5		6.2710	0.1593	39.4	
0.3		10		5.3968				39.4
	0.35		8.75		2.1298	0.0557	38.2	
0.4		7.5		3.2670				77.6
	0.45		6.75		0.9321	0.0258	36.1	
0.5		6.0		2.3349				113.7
	0.55		5.5		0.4607	0.0140	32.9	
0.6		5.0		1.8742				146.6

4-50.

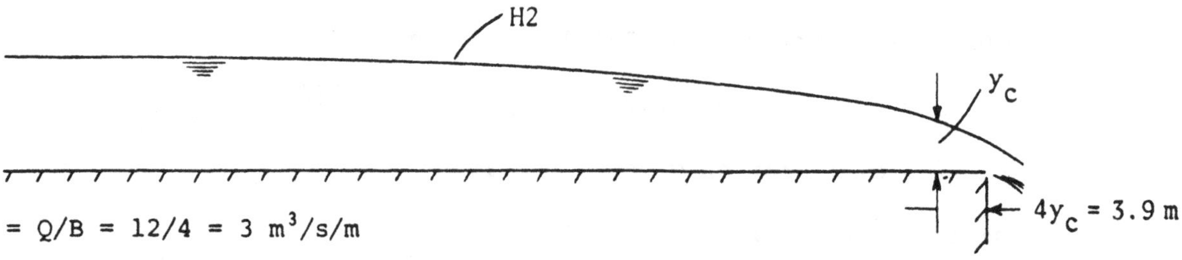

$q = Q/B = 12/4 = 3\ m^3/s/m$

$y_c = \sqrt[3]{q^2/g} = 0.972$ m (This depth occurs near brink.)

Carry out a step solution for the profile upstream from the brink.

$Re \approx V \times 4R/\nu \approx 3 \times 1/10^{-6} \approx 3 \times 10^6$; $k_s/4R \approx 0.3 \times 10^{-3}/4 \approx 0.000075$;

$f \approx 0.010$

See following page for solution table.

Check Δy from 1.2 to 1.1

$V_{avg} = Q/(A_1+A_2/2) = 12/(4.8+4.4)/2 = 2.61$ ft/s

$V_m^2 = 6.805$

$R_{1.2} = 1.2 \times 4/(6.4) = 0.75$

$R_m = 0.729$ m

$R_{1.1} = 1.1 \times 4/(6.2) = 0.7097$

$h_f/L = S_f = fV^2/(8gR_m) = (0.01 \times 6.805)/(8 \times 9.81 \times 0.729) = 1.189 \times 10^{-3}$

$\Delta h_f = S_f \times 33 = 0.0392$ m

$y_1 + V_1^2/2g = y_2 + V_2^2/2g + \Delta h_f$

$y_1 - y_2 = 0.379 - 0.319 + 0.0392 = 0.0996$ m

4-51. Upstream of jump the profile will be an H3.

Downstream of jump the profile will be an H2.

The baffle blocks will cause the depth upstream of A to increase; therefore, the jump will move towards the sluice gate.

Solution Table for Problem 4-50

Section number upstream of y_c	Depth y, m	Velocity at section V, m/s	Mean Velocity in reach $(V_1+V_0)/2$	V^2	Hydraulic Radius $R=A/P$, m	Mean Hydraulic Radius $R_m=(R_1+R_2)/2$	$S_f=fV^2_{mean}/8gR_{mean}$	$\Delta x = ((y_2+V_2^2/2g)-(y_1+V_1/2g))/S_f$	Distance upstream from brink x, m
1(at $y=y_c$)	0.972	3.086			0.654				3.9 m
			3.073	9.443		0.656	1.834×10^{-3}	0.1 m	4.0 m
2	0.98	3.060			0.658				
			3.045	9.272		0.660	1.790×10^{-3}	0.4 m	4.4 m
3	0.99	3.030			0.662				
			2.986	8.916		0.669	1.698×10^{-3}	1.7 m	6.1 m
4	1.02	2.941			0.675				
			2.886	8.327		0.684	1.551×10^{-3}	4.7 m	10.9 m
5	1.06	2.830			0.693				
			2.779	7.721		0.701	1.403×10^{-3}	7.7 m	18.6 m
6	1.10	2.727			0.710				
			2.613	6.828		0.730	1.192×10^{-3}	33.2 m	51.8 m
7	1.20	2.500			0.750				
			2.404	5.779		0.769	9.576×10^{-4}	55.3 m	107.1 m
8	1.30	2.308			0.788				
			2.225	4.951		0.806	7.83×10^{-4}	80.0 m	187.1 m
9	1.40	2.143			0.824				
			2.0715	4.291		0.841	6.501×10^{-4}	107.4 m	294.5 m
10	1.50	2.00			0.857				

The depth 300 m upstream is approximately 1.51 m

4-52. The profile might be an M profile or an S profile depending upon whether the slope is mild or steep. However, if it is a steep slope the flow would be uniform right to the brink. Check to see if M or S slope:

$Q = (1.49/n)\ A\ R^{0.667}\ S^{0.5}$

$AR^{2/3}/b^{8/3} = Q/((1.49/n)\ (S^{0.5})\ (b^{8/3}))$; assume n = 0.012

$= 120/((1.49/0.012)\ (0.0002)^{0.5}\ (10^{8/3}))$

$= 0.147$

With a value of 0.147 for $AR^{2/3}/b^{8/3}$ one finds $y/b \approx 0.40$ from Fig. 4-7. Or $y = 0.4b =$ <u>4.0 ft</u>

Then $V = Q/A = 120/40 = 3.00$ ft/sec

$F = V/\sqrt{gy} = 3.00/(\sqrt{32.2 \times 4}) = 0.26$ (subcritical)

Therefore, the water surface profile will be an <u>M2</u>

4-53. $F_r = V/\sqrt{gy}$

where $V = 10.5/2 = 5.25$ m/s

Then $F_r = 5.25/\sqrt{32.2 \times 1} = 0.93$ (supercritical)

Because the uniform flow is supercritical the water surface profile is an S type. Also, where the depth is only 0.40 m the actual depth is below both y_n and y_c; therefore, the surface profile will be an <u>S3</u>

4-54. First determine the depth upstream of the weir.

$Q = K\sqrt{2g}\ L\ H^{3/2}$; (4-38)

where $K = 0.40 + 0.05\ H/P$. By trial and error (first assume K then solve for H, etc.) solve for H yielding $H = 2.06$ ft.

Then the velocity upstream of the weir will be

$V = Q/A = 108/(406 \times 10) = 2.66$ ft/sec

$F_r = V/\sqrt{gy} = 2.66/(32.2 \times (4.06))^{0.5} = 0.23$ (subcritical)

4-54. (continued).

The Froude number just downstream of the sluice gate will next be determined:

$V = Q/A = 108/(10 \times 0.40) = 27$ ft/sec

$F_r = V/\sqrt{gy} = 27/\sqrt{32.2 \times 0.40} = 7.52$ (supercritical)

Because the flow is supercritical just downstream of the sluice gate and subcritical upstream of the weir a jump will form someplace between these two sections.

Now determine the approximate location of the jump. Let y_2 = depth downstream of the jump and assume it is approximately equal to the depth upstream of the weir ($y \approx 4.06$ ft). By trial and error (utilizing Eq. (4-27)) it can be easily shown that a depth of 0.40 ft is required to produce the given y_2. Thus the jump will start immediately downstream of the sluice gate and it will be approximately 25 ft long. Actually, because of the channel resistance y_2 will be somewhat greater than $y_2 = 4.06$ ft; therefore, the jump may be submerged against the sluice gate and the water surface profile will probably appear as shown below.

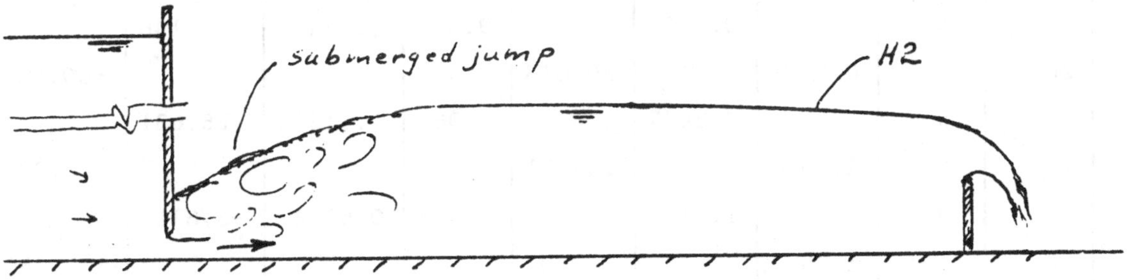

==

4-55. If part 2 is very long then a depth greater than critical will be forced in part 2 (the part with adverse slope). In that case a hydraulic jump will be formed and it may occur on part 2 or it may occur on part 1. These two possibilities are both shown below. The other possibility is for no jump to form on the adverse part. Also see this below:

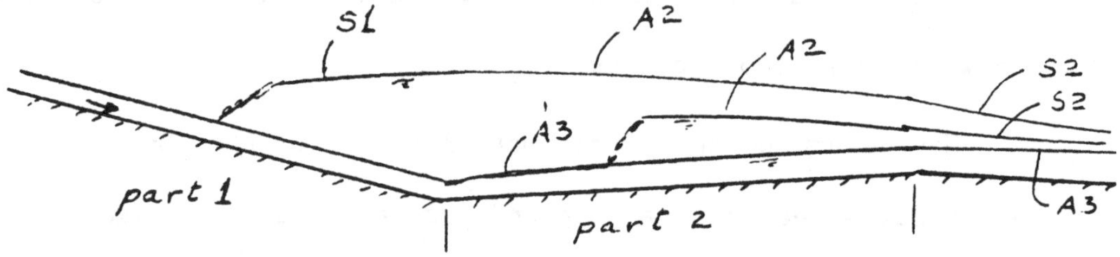

4-57. $q = 10\ m^3/s/m$ $y_c = \sqrt[3]{q^2/g} = \sqrt[3]{10^2/9.81} = 2.17\ m$

y	$\bar{y}$	V	$\bar{V}$	E	ΔE	$S_f \times 10^4$	Δx	x	elev.
52.17		0.1917		52.170				0	52.17
	51.08		0.1958		2.168	0.00287	-5,429		
50		0.20		50.002				-5,430	52.17
	45		0.2222		9.999	0.00419	-25,024		
40		0.25		40,003				-30,450	52.18
	35		0.2857		9.997	0.00892	-25,048		
30		0.333		30.006				-55,550	52.22
	25		0.400		9.993	0.02447	-25,146		
20		0.50		20.013				-80,650	52.26
	15		0.6667		9.962	0.11326	-25,631		
10		1.00		10.051				-106,280	52.51
	9		1.1111		1.971	0.5244	-5,671		
8		1.25		8.080				-111,950	52.78
	7		1.4286		1.938	1.1145	-6,716		
6		1.667		6.142				-118,670	53.47

4-58. First, one has to determine whether the uniform flow in the channel is super or subcritical. Determine y_n and then see if for this y_n the Froude number is greater or less than unity.

$Q = (1.49/n)\ AR^{2/3} S^{1/2}$; Assume $n = 0.015$

$12 = (1.49/0.015) \times y \times y^{2/3} \times (0.04)^{1/2}$

$y_n = 0.739$ ft and $V = Q/y_n = 16.23$ ft/s

$F = V/\sqrt{gy_n} = 3.33$ Therefore, uniform flow in the channel is

supercritical and one can surmise that a hydraulic jump will occur upstream of the weir. One can check this by determining what the sequent depth is. If it is less than the weir height plus head on the weir then the jump will occur.

Get sequent depth:

$$y_2 = (y_1/2)(\sqrt{1 + 8F_1^2} - 1)$$

$$= (0.739/2)(\sqrt{1 + 8 \times 3.33^2} - 1)$$

$$y_2 = \underline{\underline{3.13 \text{ ft}}}$$

Get head on weir:

$$Q = K\sqrt{2g}\ LH^{3/2} \qquad \text{Assume } K = 0.42$$

$$12 = 0.42\sqrt{64.4} \times 1 \times H^{3/2}$$

$H = 2.33$ ft; $H/P = 2.33/3 = 0.78$ so $K = 0.40 + 0.05 \times 0.78$

Better estimate for H: $H = 2.26$ ft $= 0.44$

Then depth just upstream of weir = 3 + 2.26 = $\underline{\underline{5.56 \text{ ft.}}}$
Therefore, it is proved that a jump will occur.
A rough estimate for the distance to where the jump will occur may be found by applying Eq. (4-33) with a single step computation. A more accurate calculation would include several steps.

The single-step calculation is given below:

$$\Delta x = (y_1 - y_2) + (V_1^2 - V_2^2)/2g/(S_f - S_0)$$

where $y_1 = 3.13$ ft; $V_1 = q/y_1 = 12/3.13 = 3.83$ ft/s; $V_1^2 = 14.67\ \text{ft}^2/\text{s}^2$

$y_2 = 5.56$ ft; $V_2 = 2.16$ ft/s $\qquad V_2^2 = 4.67\ \text{ft}^2/\text{s}^2$

$S_f = fV_{avg}^2/(8gR_{avg})$; $V_{avg} = 3.00$ ft/s; $\qquad R_{avg} = 4.34$ ft

Assume $k_s = 0.001$ ft; $k_s/4R = 0.00034$

$$Re = V \times 4R/\nu = ((3.83 + 2.16)/2) \times 4 \times 4.34/(1.22 \times 10^{-5}) = 4.3 \times 10^6$$

Then $f = 0.015$ and $S_f = 0.015 \times 3.0^2/(8 \times 32.2 \times 4.34) = 0.000121$

4-58. (Continued)

$\Delta x = (3.13 - 5.56) + (14.67 - 4.67)/(64.4)/(0.000121 - 0.04) = \underline{\underline{57.0 \text{ ft}}}$

Thus the water surface profile is shown below:

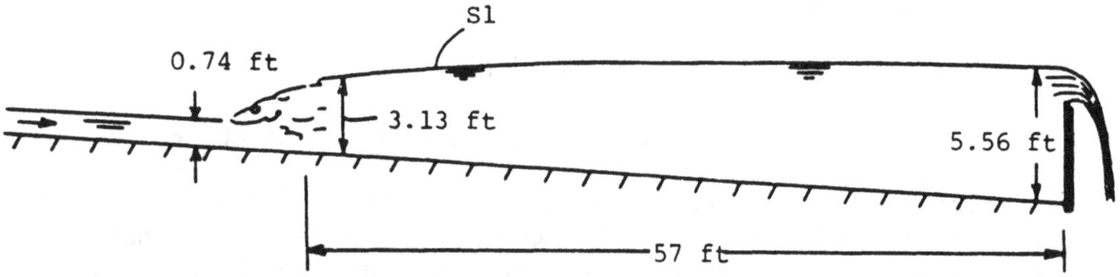

4-59. $Q = \Sigma V_i A_i$

V	A	V x A
1.32 m/s	7.6 m	10.0
1.54	21.7	33.4
1.68	18.0	30.2
1.69	33.0	55.8
1.71	24.0	41.0
1.75	39.0	68.2
1.80	42.0	75.6
1.91	39.0	74.5
1.87	37.2	69.6
1.75	30.8	53.9
1.56	18.4	28.7
1.02	8.0	8.2

$\underline{\underline{Q = 549.1 \text{ m}^3/\text{s}}}$

4-60. $Q = K\sqrt{2g}\, LH^{3/2}$ where $L = 2\text{m}$; $H = 0.13$ m; $H/P = 0.43$

Then $K = 0.40 + 0.05 \times 0.43 = 0.422$

$Q = 0.422\sqrt{2 \times 9.81} \times 2 \times (0.13)^{3/2} = \underline{\underline{0.175 \text{ m}^3/\text{s}}}$

4-61. $Q = K\sqrt{2g}\, LH^{3/2}$ where $L = 2\text{m}$, $H = 0.25\text{m}$, $H/P = 0.25$

Then $K = 0.40 + 0.25 \times 0.05 = 0.413$

So $Q = 0.413\sqrt{2 \times 9.81} \times 2 \times (0.25)^{3/2} = \underline{\underline{0.457 \text{ m}^3/\text{s}}}$

4-62. $Q = K\sqrt{2g}\, LH^{3/2}$ where $L = 6$ ft, $H = 1.5$ ft, $H/P = 0.75$

Then $K = 0.40 + 0.75 \times 0.05 = 0.4375$

$Q = 0.4375\sqrt{2 \times 32.2} \times 6 \times 1 = \underline{\underline{21.1 \text{ ft}^3/\text{s}}}$

4-63. The flow over the highway is as if flow were occurring over a broad crested weir.

$$Q = 0.385\ CL\ \sqrt{2g}\ H^{3/2}; \qquad (4\text{-}42)$$

Assume C = 1.00

$$Q = 0.385\ (100)\ \sqrt{64.4}\ (101.0 - 100.1)^{3/2}$$

$$= \underline{264\ \text{cfs}}$$

Critical depth will occur at pt B.

$y_c = (q^2/g)^{0.333}$ where q = 2.64 cfs

$y_c = y_b = (2.64^2/32.2)^{0.333} = \underline{0.600\ \text{ft}}$

4-64. $Q = 0.179\sqrt{2g}\ H^{5/2}$ where H = 1.6 ft

Then $Q = 0.179\sqrt{2g} \times (1.6)^{5/2} = \underline{4.65\ \text{ft}^3/\text{sec}}$

4-65. $Q = K\sqrt{2g}\ LH^{3/2}$ where L = 3m, Q = 6 m³/s

Assume $K \approx 0.41$ then $H = (Q/(0.41\sqrt{2g} \times 3))^{2/3}$

$H = (6/(0.41 \times \sqrt{2 \times 9.81} \times 3))^{2/3} = 1.10$ m

Then $P \approx 2.0 - 1.10 = 0.90$ m $H/P \approx 1.22$

Try again: $K = 0.40 + 1.22 \times 0.05 = 0.461$

$H = (6/(0.461 \times \sqrt{2 \times 9.81} \times 3))^{2/3} = 0.986$ m

So height of weir $P = 2.0 - 0.986 = \underline{1.01\ \text{m}}$; $H/P = 0.976$

Try again: $K = 0.40 + 0.976 \times 0.05 = 0.449$

$H = (6/(0.449 \times \sqrt{2 \times 9.81} \times 3))^{2/3} = 1.00$ m

$P = 2.00 - 1.00 = \underline{1.00\ \text{m}}$

4-66. $Q = 0.545\sqrt{g}\ LH^{3/2}$; L = 10 m and H = 0.60 m

Then $Q = 0.545\sqrt{9.81} \times 10 \times (0.60)^{3/2} = \underline{7.93\ \text{m}^3/\text{s}}$

4-67. $Q = 0.545\sqrt{g}\, LH^{3/2}$; $H = (Q/(0.545\sqrt{g}\, L))^{2/3}$

Then $H = (50/(0.545\sqrt{9.81} \times 20)^{2/3} = 1.29$ m

So, water surface elevation upstream = <u>101.29 m</u>

4-68. Solution is like that for P 4-67:

$H = (Q/(0.545\sqrt{g}\, L))^{2/3} = (1{,}500/(0.545\sqrt{32.2} \times 50))^{2/3} = 4.55$ ft

Then W.S. elevation upstream = <u>304.55 ft</u>

4-69. a) Assume the weir does not span the entire channel:

$Q = K\,(L - 0.2H)\,\sqrt{2g}\; H^{3/2}$

where $K = 0.40 + 0.05\, H/P$

First try: assume $K = 0.42$ and $H = 1$ ft

$100 = 0.42\,(15 - 0.2)\,\sqrt{64.4}\; H^{3/2}$; $H = 1.59$ ft

Second try: $K = 0.40 + .05\,(1.59/2) = 0.440$

$100 = 0.440\,(15 - 0.2 \times 1.59)\,\sqrt{64.4}\; H^{3/2}$; <u>$H = 1.550$</u>

The water surface in the channel downstream of the weir will undoubtedly be at least a foot in elevation below the weir crest; therefore, the head loss due to the weir will be $\Delta H \approx (1 + 1.55) =$ <u>2.55 ft</u>

b) $Q = 0.385\, C\, L\, \sqrt{2g}\; H^{3/2}$

Because of rounding of the upstream corner initially assume $C \approx 1.0$

First try: $100 = 0.385\,(20)\,\sqrt{64.4}\; H^{3/2}$; $H = 1.378$ ft

$H/(H + P) = 1.378/(2 + 1.378) = 0.41$

Then from Fig. 4-37, $C = 0.89$

Because of rounding of upstream corner $C = 1.03 \times 0.89 = 0.92$

Second try: $100 = 0.385\,(0.92)\,(20)\,\sqrt{2g}\; H^{3/2}$; <u>$H = 1.457$ ft</u>

Assume the water surface in the channel downstream will be at least a foot below the crest of the weir. Thus headloss, $\Delta H \approx (1 + 1.5) =$ <u>2.5 ft</u>

4.69. c) $Q = (8/15)\ K \sqrt{2g}\ \tan(\theta/2)\ H^{5/2}$, where K = 0.57

$100 = (8/15)(0.57)\sqrt{64.4}\ \tan 45°\ H^{5/2}$; $H = \underline{4.42\ ft}$

head loss: $\Delta H \approx 4.42 + 1 \approx \underline{5.4\ ft}$

d) Parshall flume W = 6 ft

$Q = K\sqrt{2g}\ W\ H_u^{3/2}$

First assume $K \approx 0.47$, then $H_u = 2.69$ ft

Second try: $H_u/B = 2.69/6.0 = 0.45$; $K = 0.49$ (Fig. 4-39)

$100 = 0.49\sqrt{2g}\ (6)\ H_u^{3/2}$; $H_u = \underline{2.62\ ft}$

Assume the head loss through the flume will be equal to 0.5 the maximum velocity head. It would be close to 1.0 $V_{max}^2/2g$ if it were not for the gradual expansion downstream of the throat section.

$V_{max} = V_{crit}$

but $y_c = (q^2/g)^{0.333}$ where $q = 100/6 = 16.67$ cfs/ft

$y_c = (16.67^2/32.2)^{0.333} = 2.05$ ft

$V_{max} = V_{crit} = Q/A_{crit} = 100/(6 \times 2.05) = 8.13$ ft/sec

$H = h_L \approx 0.5 \times 8.13^2/64.4 = \underline{0.51\ ft}$

e) Sluice gate y = 1/2 ft, L = 10 ft

$Q = KLy\sqrt{2gh}$; $V \approx 100/5 = 20$ ft/sec

assume K = 0.58 (Fig. 4-42)

$100 = 0.58 \times 10 \times 0.5\sqrt{2gH}$; $H = 18.5$ ft

$H/y \approx 37$; $K = 0.59$

try again with K = 0.59; $\underline{H = 17.9\ ft}$; $\underline{\Delta H \approx 17\ ft}$

f) Tainter gate: y = 0.5 ft, L = 10 ft

a = 5 ft, r = 10 ft

a/r = 0.50, H/r = 2

$K \approx 0.63$ (Fig. 4-42)

4-69. (f) (continued).

$Q = 0.63 \times 10 \times .5 \sqrt{2gH}$

solving for H yields $\underline{H = 15.6 \text{ ft}}$

$H/r = 1.56$, so $K = 0.63$ (Fig. 4-42), thus

the solution is OK, $\underline{\Delta H \approx 15 \text{ ft}}$

4-70.

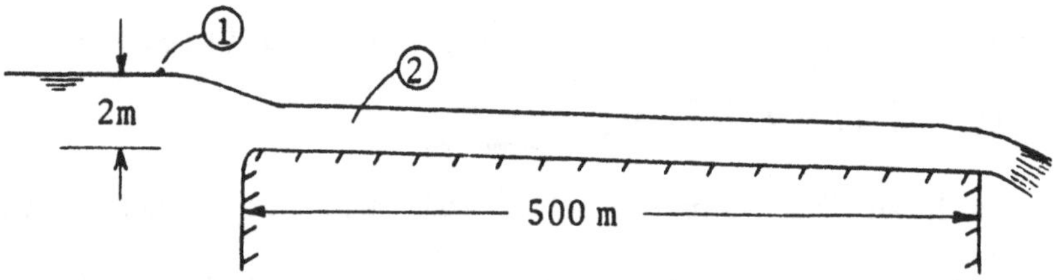

The channel is steep; therefore, critical depth will occur just inside the channel entrance. Then write the energy equation from the reservoir, (1), to the entrance section (2).

$y_1 + V_1^2/2g = y_2 + V_2^2/2g$; Assume $V_1 = 0$

Then $2 = y_2 + V_2^2/2g = y_c + 0.5\, y_c$

Solving for y_c: $y_c = 2/1.5 = 1.33$ m

Get $V_c = V_2$: $V_c^2/g = y_c = 1.33$ or $V_c = 3.62$ m/s

Then $Q = V_c A_2 = 3.62 \times 1.33 \times 4 = \underline{19.2 \text{ m}^3/\text{s}}$

4-71. a) Assume uniform flow is established in the channel except near the downstream end. Then if the energy equation is written from the reservoir to a section near the upstream end of the channel, we have:

$$2.5 \approx V_n^2/2g + y_n \quad (1)$$

$$\text{Also, } V_n = (1/n)R^{2/3} S^{1/2} \text{ or } V_n^2/2g = (1/n^2)R^{4/3} S/2g \quad (2)$$

$$\text{where } R = A/P = 3.5y_n/(2y_n + 3.5) \quad (3)$$

Then combining Eqs. (1), (2), and (3) we have

$$2.5 = ((1/n^2)((3.5y_n/(2y_n + 3.5))^{4/3} S/2g) + y_n \quad (4)$$

Assuming $n = 0.012$ and solving Eq. (4) for y_n yields:

$y_n = 2.16$ m; Also solving (2) yields $V_n = 2.58$ m/s

Then $Q = VA = 2.58 \times 3.5 \times 2.16 = \underline{19.5 \text{ m}^3/\text{s}}$

4-71.(continued).

b) With only a 100 m-long channel, uniform flow will not become established in the channel; therefore, a trial-and-error type of solution is required. Critical depth will occur just upstream of the brink so assume a value of y_c, then calculate Q and calculate the water surface profile back to the reservoir. Repeat the process for different values of y_c until a match between the reservoir water surface elevation and the computed profile is achieved.

4-72. Assume uniform flow in the channel underneath the girder.

a) $Q = (1.49/n)\ A\ R^{2/3}\ S^{1/2}$

$= (1.49/0.013)\ (40d)\ ((40d/(2d + 40)^{2/3}\ (0.0004)^{1/2}$

When the girder is in place d = 10 ft when the water surface is at the level of the bottom of the girder.

With d = 10 ft, Q = 3,288 cfs

b) Estimate flow at which the girder will be first overtopped.

The girder will act like a sluice gate (see below).

Assume submerged flow with $y_d/y \approx 1.4$

$F_o \approx V_o/\sqrt{gy_o}$

$\approx 15/\sqrt{32.2 \times 10}$

$= .84$

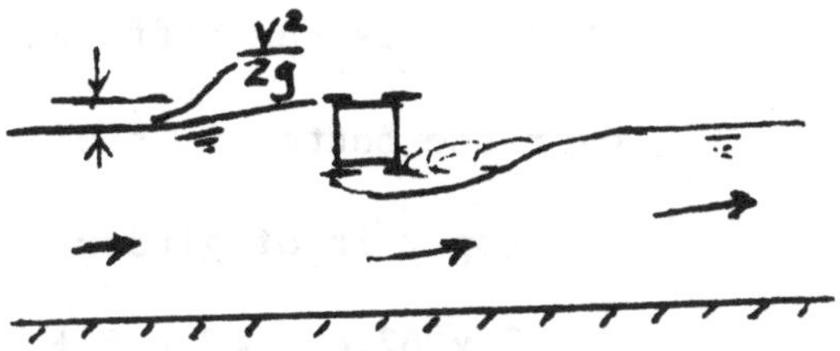

From Fig. 4-41, K ≈ 0.45

then $Q = KLy\sqrt{2gH}$

$= 0.45 \times 40 \times 10\sqrt{2g \times 18}$

$= 6,128$ cfs; $V_o = Q/A_o = 6,128/400 = 15.3$ ft/sec

We assumed $V_o = 15$ ft/sec (above for calculating F_o);

therefore, F_o and K are OK, so Q = 6,128 cfs

Next determine the uniform flow depth (downstream depth) for a discharge of 6,128 cfs

From (a) $AR^{2/3}/b^{8/3} = Q_n/(1.49\ S^{1/2}b^{8/3})$

$= (6,128)\ (0.013)/((1.49)\ (.0004)^{1/2}(40)^{8/3})$

Then from Fig. 4-7, $y/b = 0.40$, and $y = 16$ ft $= y_d$

Then better estimate of y_d/y is $16/10 = 1.6$. This is close to our initially assumed value of 1.4; therefore, K is still OK

c) The girder and water surfaces are as shown below:

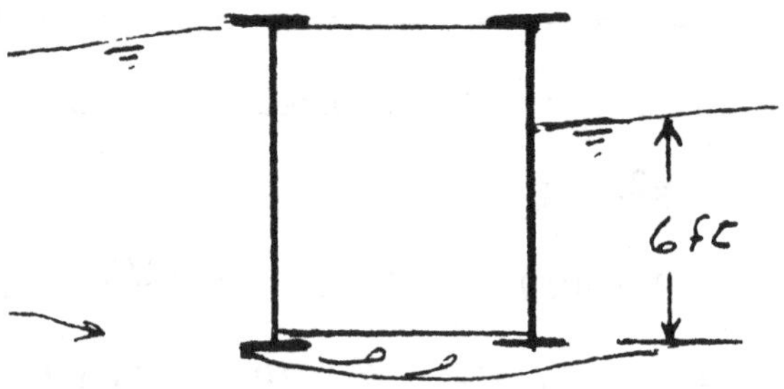

Because of the flange on the girder the pressure upstream and downstream of the girder will probably be closely approximated by a hydrostatic pressure distribution. Thus the force on the girder will be given as:

$F = \Delta p \times A_6 + \bar{p}\, A_2$

where Δp is the difference in upstream and downstream pressure acting over the bottom 6 ft of girder and $\bar{p}$ is the average pressure acting on the top 2 ft of girder.

$F = 2 \times 62.4 \times 40 \times 6 + (1/2) \times 2 \times 62.4 \times 40 \times 2 =$

$=$ <u>34,944 lbs</u>

d) $Q_{w/o\ bridge} = (1.49/0.013)\ (40 \times 18)\ ((40 \times 18)\ /$
$(36 + 40))^{2/3}\ (.0004)^{1/2}$

$=$ <u>7,390 cfs</u>

$Q_{reduction} = 7{,}390 - 6{,}128 =$ <u>1,262 cfs</u>

Note: The above solution for answers for 4-72 (b) and (c) is highly dependent on the depth downstream of the girder. In the solution given here it was assumed that uniform flow depth prevailed downstream

4-72. (continued).

of the girder. However, this downstream depth is highly sensitive to the location of the control. Therefore, other solutions might be just as valid depending upon the assumption relating to the downstream depth. For example, one could also assume free flow downstream of the girder. The essence of this problem is not necessarily one of obtaining a particular answer but of visualizing the possible ways of approaching the problem consistent with reasonable assumptions.

4-73. A detailed design is not included because each designer will make different decisions. However, the following items should be addressed in the design:

a. choice of channel cross section

b. excavation to yield relatively fixed channel slope. However, one may want two sections with different slopes for each.

c. flow will undoubtedly be supercritical; therefore, special attention would have to be given to the channel bend to prevent overtopping. This will also require a lined channel and special outlet works to prevent erosion.

d. Inlet should be carefully designed to prevent development of excessively high waves in the channel.

e. Cavitation may be a problem. Much of the above material is beyond the scope of this chapter; however, the instructor may wish to assign the problem as a means of initiating interest and discussion for these advanced subject areas.

4-74.

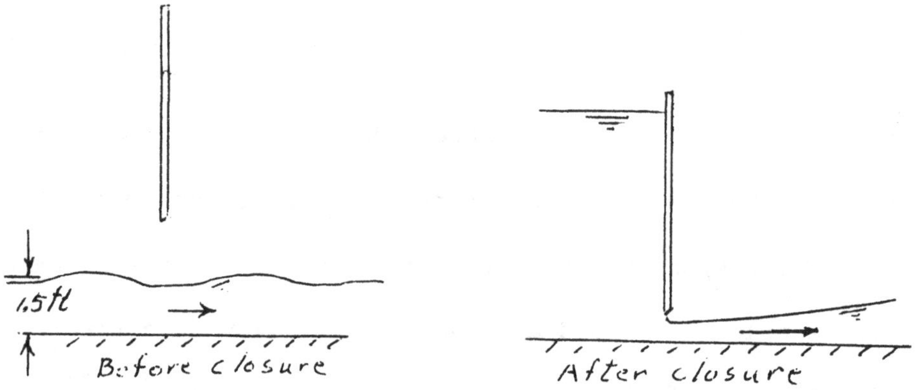

The flow conditions before and after the gate is lowered are shown above. The Froude number of the flow before the gate is lowered is now calculated.

$$Fr = V/\sqrt{gy}$$

$$V = Q/A = 100/15 = 6.67 \text{ ft/s}$$

$$Fr = 6.67/\sqrt{32.2 \times 1.5} = 0.96$$

The initial flow is very close to critical; therefore, it is obvious that when the gate is lowered the downstream flow will be supercritical and the flow upstream of the gate will be subcritical.

Assume we have free-flow conditions downstream of the gate because the flow will tend toward uniform flow which is very close to critical. From Fig. 4-42(a) it is seen that $K \approx 0.55$ and assume negligible head loss across the gate. Then the velocity downstream of the gate will be

$$V_2 = Q/A = 100/(0.55 \times 10)$$

$$= 18.18 \text{ ft/s}$$

$$Fr_2 = V_2/\sqrt{gy}$$

$$= 18.18/\sqrt{32.2 \times 0.55}$$

$$= 4.32$$

Now write the energy equation from upstream of the gate to downstream of the gate to obtain the upstream depth (neglect head loss).

$$y_1 + V_1^2/2g = y_2 + V_2^2/2g$$

$$y_1 + Q^2/(2gA_1^2) = y_2 + V_2^2/2g$$

but $A_1 = 10\, y_1$

so $y_1 + Q^2/(2g \times 100\, y_1^2) = 0.56 + 18.18^2/2g$

4-74. (continued)

$$y + 100/(64.4\, y_1^2) = 5.69$$

$$y = \underline{\mathbf{5.64\ ft}}$$

Because the uniform flow condition is very close to critical, lets call the water surface profile downstream of section (2) a C3 profile and the profile upstream of the gate a C1 profile. The sketches of these profiles are shown below.

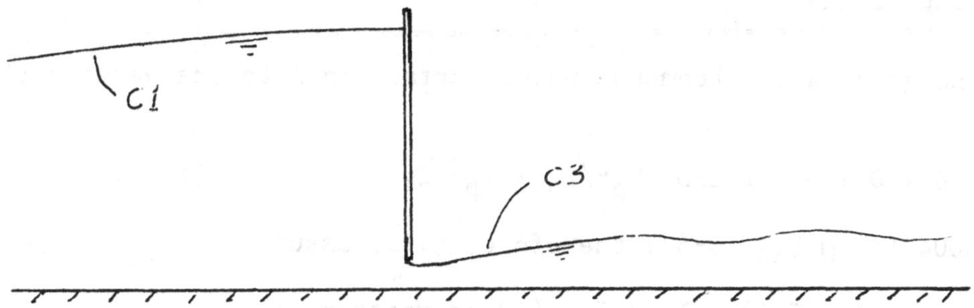

CHAPTER FIVE

5-1. a) Solid line is EGL, dashed line is HGL
b) No; AB is smallest.
c) from B to C
d) p_{max} is at the bottom of the tank
e) p_{min} is at the bend C.
f) A nozzle.
g) above atmospheric pressure
h) abrupt expansion

5-2. Write the energy equation from the water surface in A to the water surface in B.

$$0 + 0 + H = 0 + 0 + 0 + f(L/D)\, V_p^2/2g + V_p^2/2g \qquad (1)$$

$k_s/D \approx 0.00004$ (from Fig. 5-5); therefore, first assume

$f = 0.011$ (from Fig. 5-4). Then Eq. (1) is written as

$$16 = (0.011(300/1) + 1)\, V_p^2/2g$$

or $V_p = \sqrt{(16/4.3)\,2g} = 8.54$ m/s

check f: $Re = VD/\nu = 8.54 \times 1/(1.3 \times 10^{-6}) = 7 \times 10^6$

so from Fig. 5-4 $f = 0.010$

Solve for V_p again with better value for f:

$V_p = \sqrt{(16/4)\,2g} = 8.86$ m/s

$Q = VA = 8.86 \times (\pi/4) \times 1^2 = \underline{6.96\ m^3/s}$

To determine p_p write the energy equation between the water surface in A and point P:

$$0 + 0 + H = p_p/\gamma + V_p^2/2g - h + 0.01 \times (150/1) V_p^2/2g$$

$16 = p_p/\gamma - 2 + 2.5\, V_p^2/2g$ where $V_p^2/2g = 4$ m

Then $p_p = 9.810\,(16 + 2 - 10) = \underline{78.5\ kPa}$

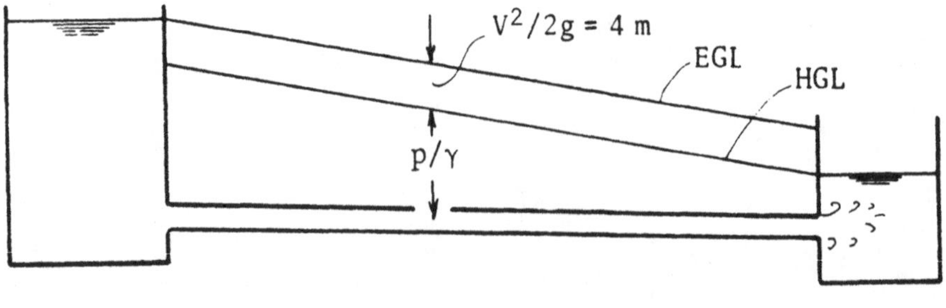

5-3. In Prob. 5-2 the discharge varies as the first power of V but V varies inversely to the 1/2 power of f:

$$V = kf^{-1/2} \text{ or } Q = k_1 f^{-1/2}$$

Then $dQ = -1/2\, kf^{-3/2}\, df$

$$dQ/Q = (-1/2\, k_1 f^{-3/2}\, df)/(k_1 f^{-1/2})$$

$$\underline{dQ/Q = (-1/2)(df/f)}$$

Thus a +10 % error in f would produce a **-5% error in Q.**

5-4. Write the energy equation from the water surface in reservoir B to the water surface in reservoir A.

$$p_B/\gamma + V_B^2/2g + z_B + h_p = p_A/\gamma + V_A^2/2g + z_A + \Sigma h_L$$

But p_B and $p_A = 0$ gage pressure so we have

$$0 + 0 + h_p = 0 + 0 + 16 + \Sigma h_L$$

where $\Sigma h_L = (K_e + fL/D + K_E)\, V^2/2g$

$$V = 3 \text{ m/s}$$

$$Re = VD/\nu = 3 \times 1/(1.3 \times 10^{-6})$$

$$Re = 2.3 \times 10^6$$

$k_s/D = 0.00005$ (from Fig. 5-5)

$f = 0.0115$ (from Fig. 5-4)

$$h_p = 16 + (0.5 + (0.0115 \times 300/1) + 1)(3^2/(2 \times 9.81))$$

$$= 16 + (2.27)$$

$$= 18.27 \text{ m}$$

Power $= Q\gamma h_p/\text{eff.}$

$= 3 \times (\pi/4)(1^2) \times 9{,}810 \times 18.27/0.75$

$=$ **563 kW**

The minimum pressure will occur on the suction side of the pump. Solve for this pressure by writing the energy equation from the water surface in reservoir B to the suction side of pump. Assume the pump is midway between A & B.

$$p_B/\gamma + V_B^2/2g + z_B = p_{min}/\gamma + V_s^2/2g + z_s + \Sigma h_L$$

$$0 + 0 + 2 = p_{min}/\gamma + V^2/2g + 0 + \Sigma h_L$$

$$p_{min}/\gamma = 2 - (1 + K_e + fL/D)\, V^2/2g$$

$$= 2 - (1 + 0.5 + 1.725)\, 0.459$$

$$= 2 - 1.48 = 0.52 \text{ m}$$

Then p_{min} = **5.10 kPa**

5-5. First estimate f and H

$$V = Q/A = 0.3/((\pi/4) \times 0.4^2) = 2.387 \text{ m/s}$$

Assume $T = 10^\circ C$ Then $\nu = 1.3 \times 10^{-6}$ m²/s and $\gamma = 9{,}810$ N/m³

$$Re = VD/\nu = 2.387 \times 0.4/1.3 \times 10^{-6} = 7 \times 10^5$$

k_s/D 0.0001 (from Fig. 5-5)

Then $f = 0.014$ (from Fig. 5-4)

Thus $H = \Sigma h_L$

$$= (1.5 + fL/D)V^2/2g$$

$$= (1.5 + 0.014 \times 4{,}000/0.4)(2.387^2)/(2 \times 9.81)$$

$$= 141.5 \times 0.290$$

= **41.1 m**

With $Q = 0.90$ m³/s; $V = 0.90/(\pi/4 \times D^2)$

and our energy equation is

$$H = (1.5 + f \times 4{,}000/D)(0.90^2/((\pi/4)^2 \times (D^4) \times 2 \times 9.81))$$

or $41.1 = (1.5 + 4{,}000\,(f/D))(0.0669/D^4)$

The above equation must be solved by trial and error. First assume $f = 0.013$

5-5. (Continued)

Then $41.1 = (1.5 + 52/D)(0.0669/D^4)$; $D = 0.62$ m

Now check f with D = 62 cm: $k_s/D = 0.00009$ (Fig. 5-5)

$V = 0.9/(\pi/4 \times 0.62^2) = 2.98$ m/s

$Re = VD/\nu = 2.98 \times 0.62/1.3 \times 10^{-6} = 1.4 \times 10^6$

$f = 0.013$

Since f with a 62 cm pipe is equal to our assumed f, the **62 cm pipe** is adequate to carry the 0.90 m³/s discharge. If the 62 cm pipe is not available commercially, then use the next size larger that is available.

5-6. The required discharge for the additional pipe will be

$Q = 0.90 - 0.3$

$= 0.60\ m^3/s$

Solve for the diameter of pipe needed as in Prob. 5-5 but with $Q = 0.60\ m^3/s$.

The equation to solve is:

$41.1 = (1.5 + 4{,}000\ f/D)(0.60^2/((\pi/4)^2\ D^4 \times 2 \times 9.81))$

Assume $f = 0.012$

$41.1 = (1.5 + 48/D)(0.0297/D^4)$

$D = 0.52$ m

Check f: $V = Q/A = 2.82$ m/s

$Re = VD/\nu = 2.82 \times 0.52/1.3 \times 10^{-6} = 1.13 \times 10^6$

$k_s/D = 0.00009$

$f \approx 0.0125$

Iterate again with $f = 0.0125$

$D = 0.52$ **OK**

Use pipe size **D = 52 cm** or the next commercial size larger.

5-7. Neglect entrance and exit losses.

Then $h_f = (fL/D)\, V^2/2g$

$k_s/D \approx 0.00015$ (Fig. 5-5)

$Re = VD/\nu$ where $\nu = 1.2 \times 10^{-5} (T = 60^\circ F)$

$V = Q/A = 5/((\pi/4) \times 1^2)$

$= 6.37$ ft/s

$Re = 6.37 \times 1/(1.2 \times 10^{-5}) = 5.3 \times 10^5$

$f = 0.015$

$h_f = (0.015 \times 5{,}280/1)(6.37^2/(64.4))$

$= 49.9$ ft

Now solve for Q for D = 11 in.

$49.9 = (0.015 \times 5{,}280/(11/12))(Q^2/(2gA^2))$

$= 79.2\, Q^2/((\pi/4)^2 \times 64.4 \times (11/12)^5)$

$49.9 = 79.2\, Q^2/(25.71)$

$Q^2 = 16.20$ ft^6/s^2

Q = **4.02 cfs**

Solve for Q for D = 11 in. and f = 0.015 (1.5)

$49.9 = 4.62\, Q^2$

Q = 3.29 cfs

5-8. $V = Q/A = 2.5/((\pi/4) \times (2/3)^2) = 7.16$ ft/sec

$R_e = VD/\nu = 7.16 \times (2/3)/(1.1 \times 10^{-5}) = 4.3 \times 10^5$

$k_s/D = 0.0002$ (from Fig. 5-5)

Then f = 0.016 (from Fig. 5-4)

Assume entrance loss coefficient = 0.5 and exit loss coefficient = 1.0

5-8. (continued).

Then $h_L = (f(L/D) + 1.5)\, V^2/2g$

$h_L = (0.016\ (3{,}000/(2/3)) + 1.5)\ (7.16^2/64.4) = 58.5$ ft.

Write energy equation from water surface to water surface:

$p_1/\gamma + V_1^2/2g + z_1 + h_p = p_2/\gamma + V_2^2/2g + z_2 + h_L$

$0 + 0 + 100 + h_p = 0 + 0 + 150 + 58.5;\quad h_p = 108.5$ ft

Power supplied $= Q\gamma h_p = 2.5 \times 62.4 \times 108.5 = 16{,}926$ ft-lb/sec.

= 30.8 horsepower

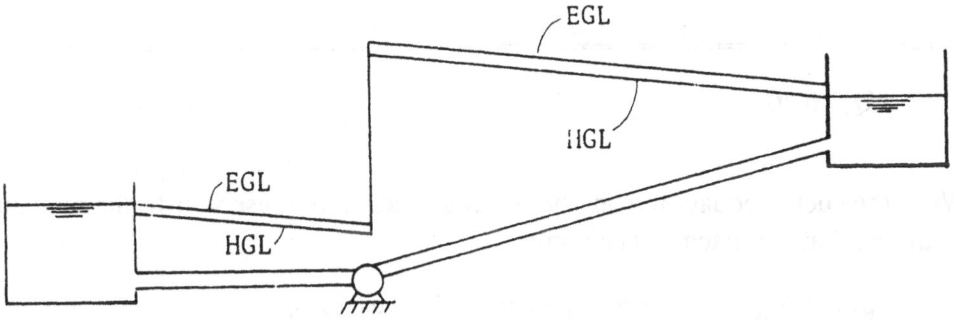

5-9. Write the energy equation from the water surface in the reservoir to the outlet assuming that the machine is a pump.

$p_1/\gamma + V_1^2/2g + z_1 + h_p = p_2/\gamma + V_2^2/2g + z_2 + \Sigma h_L$

$0 + 100 + h_p = 0 + V^2/2g + 0 + (K_e + fL/D)V^2/2g$

$h_p = (1 + 0.5 + fL/D)\ V^2/2g - 100$

$V = Q/A = 20/\ ((\pi/4) \times 1^2)$

$= 25.46$ ft/s

Compute Re: $Re = VD/\nu$

$= 25.5 \times 1/(1.22 \times 10^{-5})$(Assume T = 60°F)

$= 2.1 \times 10^6$

$k_s/D = 0.00015$ (Fig. 5-5)

$f = 0.0135$ (Fig. 5-4)

Then $h_p = (1 + 0.5 + 0.0135 \times 1{,}000/1)(25.46^2/64.4) - 100$

$h = 51$ ft

The machine is a **pump** because h_p is positive.

Assuming 100% efficiency $P = Q\gamma h_p$

$= 20 \times 62.4 \times 51$

$P = 63{,}648$ ft-lb/s

$P =$ **116 horsepower**

5-10. $Q = 6$ cfs

Write the energy equation from the water surface in the reservoir to the pipe outlet assuming that the machine is a pump.

$$p_1/\gamma + V_1/2g + z_1 + h_p = p_2/\gamma + V_2/2g + z_2 + \Sigma h_L$$

$$0 + 0 + 100 + h_p = 0 + V^2/2g + 0 + (K_e + (fL/D))V^2/2g$$

$$100 + h_p = (1 + K_e + fL/D)V^2/2g$$

Assume k_s/D 0.00015 (Fig. 5-5)

Assume $\nu = 1.22 \times 10^{-5}$ ($T = 60^\circ$ F)

$V = Q/A = 7.64$ ft/s

Then $Re = VD/\nu = 7.64 \times 1/(1.22 \times 10^{-5}) = 6.3 \times 10^5$

$f = 0.015$

Now solve for h_p

$h_p = -100 + (1 + 0.5 + (0.015 \times 1{,}000/1))\ 7.64^2/64.4$

$= -100 + (16.5)(0.906)$

$= -100 + 14.95$

$= -85.0$ ft

Since h_p turns out to be negative it is obvious that the machine is a **turbine** rather than a pump. Assuming 100% efficiency:

5-10. (continued)

$$P = Q\gamma h_p = 6 \times 62.4 \times 85.0$$

$$= 31{,}824 \text{ ft-lbs/s}$$

$$= \underline{\textbf{57.9 horsepower}}$$

5-11. $V = Q/A = 2.50/((\pi/4) \times 1^2) = 3.183$ m/s

$Re = VD/\nu = 3.183 \times 1/(10^{-6}) = 3.18 \times 10^6$

$k_s/D = 0.00005$ (Fig. 5-5) and $f = 0.0115$ (Fig. 5-4)

Now write the energy equation from river to canal:

$$p_1/\gamma + V_1^2/2g + z_1 + h_p = p_2/\gamma + V_2^2/2g + z_2 + h_L$$

$$0 + 0 + 100 + h_p = 0 + 0 + 150 + f(L/D)V^2/2g$$

$$h_p = 50 + 0.0115 \times (2{,}000/1) \times 3.183^2/(2 \times 9.81) = 50 + 11.9 = 61.9 \text{ m}$$

Then $P = Q\gamma h_p/0.82 = 2.50 \times 9{,}790 \times 61.9/0.82 = \underline{\underline{1{,}847 \text{ kW}}}$

5-12. $k_s/D = 0.00016$ (from Fig. 5-5)

Assume $f = 0.015$ (from Fig. 5-4)

Write energy equation from reservoir water surface to jet:

$$p_1/\gamma + V_1^2/2g + z_1 = p_2/\gamma + V_2^2/2g + z_2 + h_L$$

$$0 + 0 + 100 = 0 + V_2^2/2g + 60 + f\,(L/D)\,V_p^2/2g + 0.1\,V_p^2/2g$$

The head loss coefficient for the pipe entrance is assumed to have a value of 0.10. Neglect the head loss in the nozzle.

$V_2A_2 = V_pA_p$; $\quad V_p = V_2A_2/A_p$

$V_p = V_2(d_2^2/d_p^2) = (1/4)\,V_2$

Then $100 - 60 = (V_2^2/2g)\,(1 + ((1/16)\,f(L/D) + (1/16)\,(0.1))$

$40 = (V_2^2/2g)\,(1 + 0.94 + 0.01)$; $\quad V_2 = 36.35$ ft/sec

check f: $Re = VD/\nu = (36.35/4) \times 1/(1.2 \times 10^{-5}) = 7.6 \times 10^5$

$f \approx 0.015$ (from Fig. 5-4); $\quad f$ is OK

5-12. (continued)

Thus $Q = V_2 A_2 = 36.35\ (\pi/4)\ (0.5^2) = \underline{7.14\ \text{cfs}}$

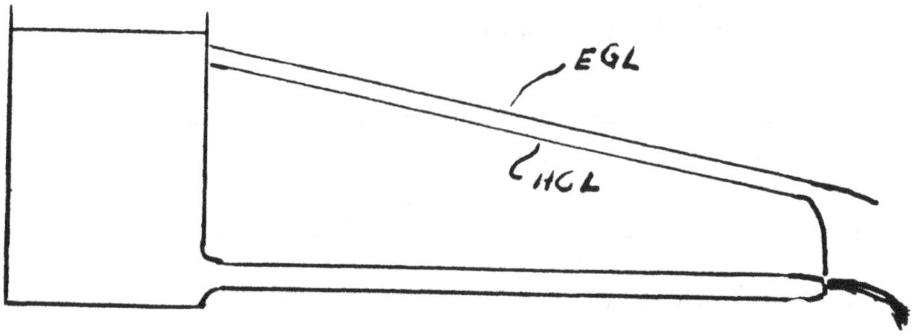

5-13. Write the energy equation from the water surface in the reservoir to the outlet.

$$p_1/\gamma + V_1^2/2g + z_1 + h_p = p_2/\gamma + V_2^2/2g + z_2 + h_L$$

$$0 + 0 + 100 + h_p = 0 + V_2^2/2g + 60 + (K_e + fL/D)\ V^2/2g$$

Assume $K_e = 0.10$

$k_s/D \approx 0.00015$ (Fig. 5-5)

$f = 0.015$ (Fig. 5-4)

$V = Q/A = Q/((\pi/4)(0.5^2))$

$= 5.093\ Q$

$V_{pipe} = Q/A_p = Q/((\pi/4) \times 1^2)$

$= 1.273\ Q$

Then the energy equation can be written as

$$hp = -40 + (5.093Q)^2/64.4 + (0.1 + 0.015 \times 1{,}000/1)(1.273Q)^2/64.4$$

$$hp = -40 + (0.403 + 0.380)\ Q^2$$

$$= -40 + 0.783\ Q^2$$

Solving for h_p for different values of Q and plotting Q vs h_p on Fig. 5-10, it is found that the system and pump curves intersect at approximately

$\underline{Q = 13.2\ \text{cfs}}$ where $h_p = 96.4$ ft.

$$\text{Power} = Q\gamma h_p/\text{eff}$$

$$= 13.2 \times 62.4 \times 96.4 / 0.70$$

$$= \underline{\mathbf{113{,}470\ ft{\cdot}lb/s}}$$

$$= \underline{\mathbf{206\ horsepower}}$$

Therefore $V_p = Q/H = 13.2 / ((\pi/4) \times 1^2)$

$V_p = 16.8$ ft/s; $V_p^2/2g = 4.39$ ft.

$Re = 16.8 \times 1/(1.22 \times 10^{-5}) = 1.4 \times 10^6$: $f = 0.015$

Pipe head loss $= (fL/D)\ V^2/2g = (0.015 \times 1{,}000/1)(4.39) = 65.9$ ft

$V_2 = 5.093\ Q = 67.2$ ft/s

$V_2^2/2g = 70.2$ ft

Upon plotting the E.G.L. and H.G.L. (see below) it is shown that the greatest pressure occurs just downstream of the pump and minimum pressure occurs just upstream of the pump.

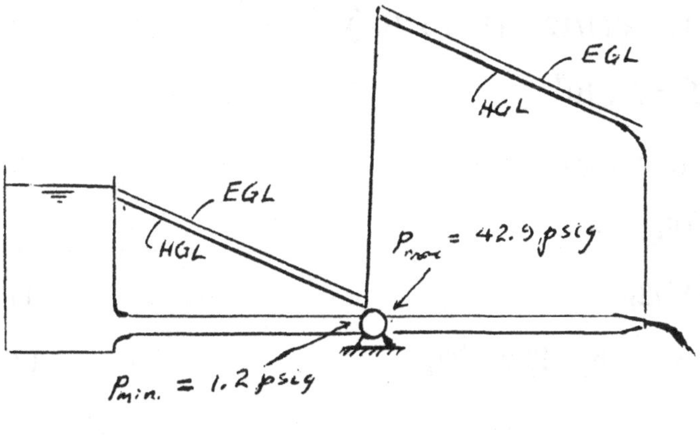

$$h_{min} = 40 - V_p^2/2g - 65.9/2$$

$$= 40 - 4.39 - 65.9/2 = 2.7 \text{ ft}$$

$$p_{min} = \gamma h_{min}$$

$$= \underline{\mathbf{168\ psfg}}$$

$$= \underline{\mathbf{1.2\ psig}}$$

5-13. (Continued)

$h_{max} = h_{min} + 96.4$

$= 2.7 + 96.4$

$= 99.1$ ft

$p_{max} = \gamma h_{max}$

$=$ **<u>6,184 psfg</u>**

$=$ **<u>42.9 psig</u>**

5-14. $Q = 25$ gpm

$= 0.0557$ cfs

$V = Q/A$

$= 0.0557 / ((\pi/4) \times (1/12)^2)$

$= 10.21$ ft/s

Assume $T = 60^\circ F$ $\nu = 1.22 \times 10^{-5}$

Then $Re = VD/\nu = 10.21 \times (1/12) / (1.22 \times 10^{-5})$

$= 0.7 \times 10^5 = 7 \times 10^4$

$k_s/D = 0.005$ (Fig. 5-5)

$f = 0.032$ (Fig. 5-4)

$h = (fL/D) V^2/2g$ (1)

Head loss in open globe valve: $h_L = 10.0\ V^2/2g$ (2)

Equate (1) and (2)

$fL/D\ V^2/2g = 10\ V^2/2g$

$L/D = 10/f$

$= 10/0.032$

<u>L/D = 312</u>

5-15. $h_f = \Delta(p/\gamma + z) = (-20 \times 144/62.4) + 30 = -16.2$ ft

Therefore, flow is from B to A

$Re\, f^{1/2} = (D^{3/2}/\nu)(2gh_f/L)^{1/2} = (2^{3/2}/(1.41 \times 10^{-5})$

$\times (64.4 \times 16.2/(3 \times 5{,}280))^{1/2} = 5.14 \times 10^4$

$k_s/D = 0.0004$ then $f = 0.0175$

$V = \sqrt{h_f 2gD/fL} = \sqrt{16.2 \times 64.4 \times 2)(0.0175 \times 3 \times 5{,}280)} = 2.74$ ft/s

$Q = VA = 2.74 \times (\pi/4) \times 2^2 =$ 8.60 cfs

5-16. Darcy-Weisbach: $k_s/D = 0.00016$ (Fig. 5-5)

$V = Q/A = 5/((\pi/4) \times 1^2) = 6.37$ ft/sec.

Assume $\nu = 1.2 \times 10^{-5}$ ft^2/sec.; $Re = VD/\nu = 6.37 \times 1/(1.2 \times 10^{-5})$

$= 5.3 \times 10^5$

Then from Fig. 5-4 $f = 0.015$

$h_f = f\,(L/D)\,V^2/2g =$ 28.3 ft.

Hazen-Williams:

Assume $C_h = 140$

Then $h_f = 3.02 \times 3{,}000 \times 1^{-1.167}\,(6.37/140)^{1.85} =$ 29.8 ft.

$h_{f(HW)}/h_{f(DW)} = 29.8/28.3 =$ 1.05

The comparison is pretty good considering the uncertainties involved in choosing k_s and C_h.

5-17. Darcy-Weisbach:

$V = Q/A = 150/((\pi/4) \times 6^2) = 5.305$ ft/sec.

Assume $\nu = 1.2 \times 10^{-5}$ ft^2/sec; so $Re = 5.305 \times 6/(1.2 \times 10^{-5}) = 2.7 \times 10^6$

$k_s/D = 0.000025$ (Fig. 5-5) and $f = 0.0105$ (Fig. 5-4)

Then $h_f = 0.0105\,(10{,}000/6)\,(5.305^2/64.4) =$ 7.65 ft.

5-17. (continued)

Hazen-William: Assume C = 140

h_f = 3.02 x 10,000 x $6^{-1.167}(5.305/140)^{1.85}$ = 8.75 ft.

Swamee-Jain: $f = 0.25/(\log((k_s/3.7D) + (5.74/Re^{0.9})))^2$

f = 0.0109

h_f = 0.0109/0.0105 x 7.65 = 7.94 ft.

5-18. Darcy-Weisbach: ν = 1.2 x $10^{-5}ft^2/sec$.

$k_s/D = 6 \times 10^{-6}$ (from Fig. 5-5 assuming degree of smoothness is halfway between steel pipe and drawn tubing)

$Re = VD/\nu$ = 12 x 5/(1.2 x 10^{-5}) = 5.0 x 10^6

Then f = 0.009

$h_f = f(L/D)(V^2/2g)$ = .009 x (1,000/5) (12^2/64.4) = 4.02 ft.

Mannings: $V = (1.49/n) R^{2/3} S^{1/2} = (1.49/n) R^{2/3} (h_f/L)^{1/2}$

$R = A/P = (\pi D^2/4)/(\pi D) = D/4$ = 1.25 ft.; $R^{2/3}$ = 1.16

Assume n = 0.010

Then 12 = (1.49/0.010) (1.16) $(h_f/1{,}000)^{1/2}$

Solving: h_f = 4.82 ft.

Hazen-Williams: Assume C = 140

$h_f = 3.02\ LD^{-1.167}(V/C_h)^{1.85}$

= 3.02 x 1000 x $5^{-1.167}(12/140)^{1.85}$ = 4.90 ft.

5-19. Darcy-Weisbach: ν = 1.2 x 10^{-5} ft^2/sec

k_s/D = 0.0002 (Fig. 5-5); Re = 5.0 x 10^6; f = 0.014 (Fig. 5-4)

$h_f = f(L/D)(V_2/2g)$ = 0.014 X (1,000/5) (12^2 /64.4) = 6.26 ft.

Mannings: Assume n = 0.011

h_f = 4.82 x $(0.011/0.010)^2$ = 5.83 ft.

5-19. (continued)

Hazen-Williams: Assume C = 120 (Table 5-2)

$h_f = 4.90\ (140/120)^{1.85} = \underline{6.52 \text{ ft.}}$

Swamee-Jain:

$f = 0.25/(\log((k_s/3.7D) + (5.74/Re^{0.9})))^2$

$= 0.25/(\log\ (5.94 \times 10^{-5}))^2 = 0.0140$

$h_f = (0.0140/0.014)(6.26) = \underline{6.26 \text{ ft.}}$

5-20. $p_1/\gamma + V_1^2/2g + z_1 = p_2/\gamma + V_2^2/2g + z_2 + h_L$

$0 + 0 + (10 + 2 \times 2 \times 5.28) = 0 + V^2/2g + 0 + f(L/D)V^2/2g$

$31.12 = V^2/2g(1 + f \times 2 \times 5{,}280/2)$

$V^2 = 2g \times 31.12/(1 + 5{,}280f) = 2{,}004/(1 + 5{,}280f)$

$k_s/D = 0.00007$; Assume f = 0.013 (Fig. 5-5)

Then $V = (2{,}004/(1 + 5{,}280 \times 0.013))^{1/2} = 5.36$ ft/sec

Then $Re = 5.36 \times 2/(1.22 \times 10^{-5}) = 8.8 \times 10^5$, so f = 0.013

$Q = VA = 5.36 \times (\pi/4) \times 2^2 = \underline{\underline{16.84 \text{ cfs}}}$

5-21. $p_1/\gamma + V_1^2/2g + z_1 = p_2/\gamma + V_2^2/2g + z_2 + \Sigma h_L$

$0 + 0 + z_1 = 0 + 0 + z_2 + \Sigma h_L$

$20 = (V^2/2g)(K_e + f(L/D) + K_0) = V^2/2g(0.5 + f(L/D) + 1.0)$

$[(\pi/4)^2 \times 2g \times 20/Q^2] = [1.5 + f(L/D)] \times D^{-4}$; $7.94\ D^4 = (1.5 + fL/D)$

For first trial assume f = 0.02 and neglect minor losses:

$D^5 = 0.02 \times 2 \times 5{,}280/7.94$; D = 1.93 ft.

Then $V = Q/A = 10/((\pi/4) \times (1.93)^2) = 3.43$ ft/sec

and $Re = 3.43 \times 1.93/(1.2 \times 10^{-5}) = 5.5 \times 10^5$; $k_s/D = 0.005$; f = 0.0175

2nd trial: $D^5 = 0.0175 \times 2 \times 5{,}280/7.94$; D = 1.88 ft = $\underline{\underline{22.5 \text{ in.}}}$

Use next commercial size larger; $\underline{\underline{D = 24 \text{ in.}}}$

5-22. Assume steel pipe will be used. Also, assume a water temperature of about 60° F or $\nu = 1.2 \times 10^{-5}$. Neglect inlet and outlet losses.

Solve Eq. (5-19) for D using $k_s = 0.00015$ ft. That solution yields

D = 1.95 ft = 23.4 in. However, the next commercial size larger is

D = 24 in. Use **D = 24 in. = 2 ft.** Compute head loss for D = 24 in. and check to see if the 24 in. size is adequate. With D = 24 in. we compute the following quantities: V = 4.77 ft/s, Re = 8×10^5, $k_s/D = 7.5 \times 10^{-5}$,

f = 0.013

Thus $h_f = (fL/D)(V^2/2g)$

$= (0.013 \times 2 \times 5{,}280 / 2)(4.77^2 / 64.4$

$= 24.2$ ft < 30 OK

5-23. Write the energy equation from the upper reservoir to the lower one.

$p_1/\gamma + V_1^2/2g + z_1 = p_2/\gamma + V_2^2/2g + z_2 + \Sigma h_L + h_T$

$0 + 0 + 200 = 0 + 0 + 100 + (K_e + K_E + f(L/D)\ (V^2/2g) + h_T$

$V = Q/A = 32/((\pi/4)\ (2^2)) = 10.19$ ft/sec.

$Re = VD/\nu = 10.19 \times 2/(1.2 \times 10^{-5}) = 1.7 \times 10^6$

$k_s/D = 0.00008$ (Fig. 5-5); $f = 0.0125$ (Fig. 5-4)

Assume Ke = 0.5 and $K_e = 1.0$

Then $h_T = 200 - 100 - (1.5 + (0.125 \times 500/2)\ (10.19^2/64.4))$

$= 92.6$ ft.

Power = $Q\gamma h_T/550 = 32 \times 62.4 \times 92.6/550 = 336$ horsepower

Power output = 336×0.85 = 286 horsepower

5-24. $p_1/\gamma + V_1^2/2g + z_1 = p_2/\gamma + V_2^2/2g + z_2 + h_f$

$(300{,}000/9{,}810) + 0 = (60{,}000/9{,}810) + 10 + h_f;\ h_f = 14.46$ m

$f(L/D)(Q^2/A^2)/2g = 14.46$

$f(L/D)[Q^2/((\pi/4)D^2)^2 2g] = 14.46$

$(4^2 fLQ^2/\pi^2)/2gD^5 = 14.46;\quad D = [(8/14.46)fLQ^2/(\pi^2 g)]^{1/5}$

Assume $f = 0.020$; Then

$D = [(8/14.46) \times 0.02 \times 140 \times (0.025)^2/(\pi^2 \times 9.81)]^{1/5} = 0.100$

Then $k_s/D = 0.002$ and $f = 0.024$; Try again:

$D = 0.100 \times (0.024/0.020)^{1/5} = \underline{0.104\text{ m}}$

Use a 12-cm pipe

5-25. $p_1/\gamma + z_1 + V_1^2/2g = p_2/\gamma + z_2 + V_2^2/2g + \Sigma h_L$

$11 = \Sigma h_L = (V_1^2/2g)(Ke + 3K_{b1} + f_1 \times 45/1) + V_2^2/2g)(K_c + 2K_{b2} + K_0 + f_2 \times 30/(1/2))$

$K_e = 0.5$; $K_0 = 1.0$; From Table 5-3, $K_{b1} = 0.35$; $K_{b2} = 0.16$; $K_c = 0.39$

Assume $f_1 = 0.015$; $f_2 = 0.016$

$11 \times 2g = V_1^2(0.5 + 3 \times 0.35 + 0.015(45)) + V_2^2(0.39 + 2 \times 0.16 + 1.0 + 0.016(60))$

$708 = V_1^2(2.23) + V_2^2(2.67) = Q^2(2.23/((\pi/4)^2(1)^4) + 2.67/((\pi/4)^2(1/2)^4)) = Q^2(72.9)$

$Q^2 = 708/72.9 = 9.71;\quad Q = 3.12$ cfs

$Re = 4Q/(\pi D\nu)$; $Re_1 = 4(3.12)/(\pi(1.22 \times 10^{-5})) = 3.3 \times 10^5$

$k_s/D = 0.00015$; $f = 0.016$; $Re_2 = 6.5\ 10^5$; $k_s/D = 0.0003$; $f = 0.016$

So $\underline{Q = 3.1\text{ cfs}}$

5-26. Write the energy equation from the water surface of the lower to the water surface of the upper reservoir.

$$p_1/\gamma + V_1^2/2g + z_1 + h_p = p_2/\gamma + V_2^2/2g + z_2 + \Sigma h_L$$

Assume flow first enters the 6-in. pipe and then into the 1-ft pipe. Therefore, we must consider entrance loss to the 6-in. pipe and expansion from the 6-in. to 1-ft pipe.

5-26. (continued)

Let V_6 = velocity in the 6-in. pipe

V_{12} = velocity in the 1-ft. pipe.

The energy equation then becomes

$$h_p = 11 + \Sigma h_L$$

$$h_p = 11 + (V_6^2/2g)(K_e + 2K_{b6} + K_{E_{6-12}} + fL/D)$$
$$+ (V_{12}^2/2g)(K_{E_{12}} + 3K_{b_{12}} + fL/D)$$

$$K_e = 0.5, \quad K_{E_{6-12}} = 0.6, \quad K_{b6} = 0.16, \quad K_{b_{12}} = 0.35$$

Assume $f_6 = 0.016$, $f_{12} = 0.015$

Then $$h_p = 11 + (V_6^2/2g)(0.5 + 2 \times 0.16 + 0.6 + 0.016 \times 30/0.5)$$
$$+ (V_{12}^2/2g)(1.0 + 3 \times 0.35 + 0.015 \times 45/1)$$

$$h_p = 11 + 2.38\ V_6^2/2g + 2.725\ V_{12}^2/2g \qquad (1)$$

$$Q_6 = Q_{12}$$

$$V_6 A_6 = V_{12} A_{12}$$

$$V_6 = V_{12} \times A_{12}/A_6$$
$$= V_{12} \times 4$$

Eq. (1) becomes

$$h_p = 11 + 2.38 \times 16\ V_{12}^2/2g + 2.725\ V_{12}^2/2g$$
$$= 11 + (V_{12}^2/2g)(40.805)$$
$$= 11 + (Q^2/2g\ A_{12}^2)(40.805)\ ;\ A_{12}^2 = (\pi/4)^2 = 0.617\ \text{ft}^4$$

$$h_p = 11 + 1.027\ Q^2 \qquad (2)$$

Solving the above system equation with the pump curve of Fig. 5-10 yields

$Q \approx 10.4$ cfs and $h_p = 125$ ft

Now check f: $V_6 = Q/A = 53.48$ ft/s

$V_{12} = Q/A = 13.37$ ft/s

5-26. (continued)

Then $Re_6 = V_6 D_6/\nu = 53.5 \times 0.5 / (1.2 \times 10^{-5})$

$= 2.22 \times 10^6$; $f = 0.015$ (Fig. 5-4)

$Re_{12} = 13.4 \times 1/(1.2 \times 10^{-5})$

$= 1.1 \times 10^6$; $f = 0.014$ (Fig. 5-4)

With these new f values the energy equation now is written as

$$h_p = 11 + (V_6^2/2g)(1.42 + 0.9) + (V_{12}^2/2g)(2.05 + 0.63)$$

$$= 11 + (V_6^2/2g)(2.32) + (V_{12}^2/2g)(2.68)$$

$$= 11 + 16 \times 2.32\, V_{12}^2/2g + 2.68\, V_{12}^2/2g$$

$$= 11 + V_{12}^2/2g\,(39.8)$$

$$h_p = 11 + 1.00\, Q^2 \qquad (3)$$

Solving Eq. (3) with the pump curve of Fig. 5-10 yields

Q 10.5 cfs hp 120 ft

Power = $Q\gamma h_p$/eff.

$= 10.5 \times 62.4 \times 120/0.75$

$= 104{,}830$ ft-lb/s

= **191 horsepower**

5-27. $k_s/D = 0.004$ Assume $f = 0.028$ and $r/d \approx 2 \rightarrow K_b \simeq 0.2$

a)

$$p_1/\gamma + z_1 + V_1^2/2g = p_2/\gamma + z_2 + V_2^2/2g + \Sigma h_L$$

$$100 = 64 + (V^2/2g)(1 + 0.5 + K_b + f \times L/D)$$

$$= 64 + (V^2/2g)(1 + 0.5 + 0.2 + 0.028 \times 100/1)$$

$$36 = (V^2/2g)(4.5)$$

$$V^2 = 72g/4.5 = 515 \rightarrow V = 22.7$$

$Re = 22.7(1)/(1.22 \times 10^{-5}) = 1.9 \times 10^6$; $f = 0.028$

$Q = 22.7(\pi/4)\, 1^2$ 17.8 cfs

5-27. (continued).

b) $V^2/2g = 36/4.5 = 8.0$ ft

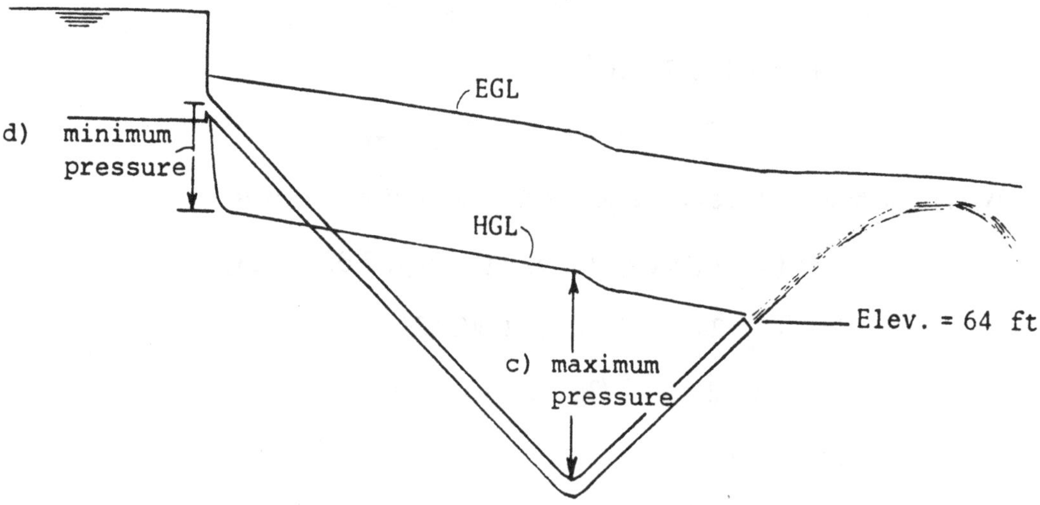

e) $P_{min}/\gamma = 100 - 95 - (V^2/2g)(1 + 0.5) = 5 - 8(1.5) = -7$ ft

$P_{min} = -7(62.4) = -437$ psfg = $\underline{-3.03 \text{ psig}}$

$P_{max}/\gamma + V_m^2/2g + z_m = P_2/\gamma + z_2 + V_2^2/2g + \Sigma h_L$

$P_{max}/\gamma = 64 - 44 + 8.0(0.2 + 0.028(28/1)) = 27.9$ ft

$P_{max} = 27.9(62.4) = 1{,}739$ psfg = $\underline{12.1 \text{ psig}}$

5-28. The solution to Prob. 5-27 shows that the discharge without the pump will be 17.8 cfs. However, the pump of Fig. 5-10 does not even produce a positive head with Q = 17.8 cfs; therefore, that pump would not augment the flow. It would probably decrease the flow somewhat. **This would not be a good installation!**

5-29. One possible design given below:

$L \approx 300 + 50 + 50 = 400$ m; $K_b = 0.19$

$50 = \Sigma h_L = V^2/2g(K_e + 2K_b + f(L/D) + 1.0) = V^2/2g(1.88 + f(L/D))$

$50 = [Q^2/(2gA^2)](f(L/D) + 1.88) = [2.5^2/(2 \times 9.81 \times A^2)]((400\ f/D) + 1.88)$

Assume f = 0.015. Then $50 = [0.318/((\pi/4)^2 \times D^4)](0.015 \times (400/D)) + 1.88)$

Solving, one gets $D \approx 0.59$ m = 59 cm. Try commercial size D = 60 cm.

5-29. (continued).

Then $V_{60} = 2.5/((\pi/4) \times 0.6^2) = 8.84$ m/s

$Re = 8.8 \times 0.6/10^{-6} = 5.3 \times 10^6$; $k_s/D = 0.0001$ and $f \approx 0.013$

Since f = 0.13 is less than originally assumed f, the design is conservative. So use D = 60 cm and L ≈ 400 m.

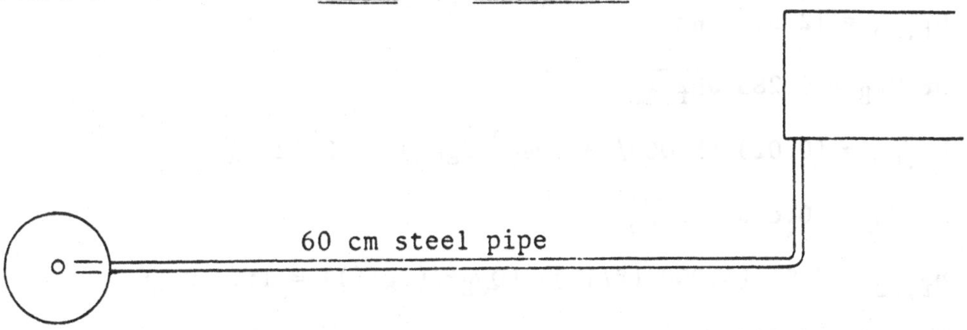

5-31. $p_1/\gamma + V_1^2/2g + z_1 + h_p = p_2/\gamma + V_2^2/2g + z_2 + \Sigma h_L$

$0 + 0 + 10 + h_p = 0 + 0 + 20 + V_2^2/2g(K_e + fL/D + k_0$

$h_p = 10 + (Q^2/(2gA^2))(0.1 + 0.02 \times 1{,}000/(10/12) + 1)$

$A = (\pi/4) \times (10/12)^2 = 0.545 \text{ ft}^2$

$h_p = 10 + 1.31\, Q_{cfs}^2$ but $Q_{cfs} = 449$ gpm

$h_p = 10 + 1.31\, Q_{gpm}^2/(449)^2$

$h_p = 10 + 6.51 \times 10^{-6} Q_{gpm}^2$

Q →	1,000	2,000	3,000
h →	16.5	36.0	68.6

Plotting this on pump curve figure yields Q ~ 2,950 gpm

5-32. An iterative solution yields the following distribution of flow (to the nearest 0.1 cfs) and pressure (to the nearest 0.1 psi)

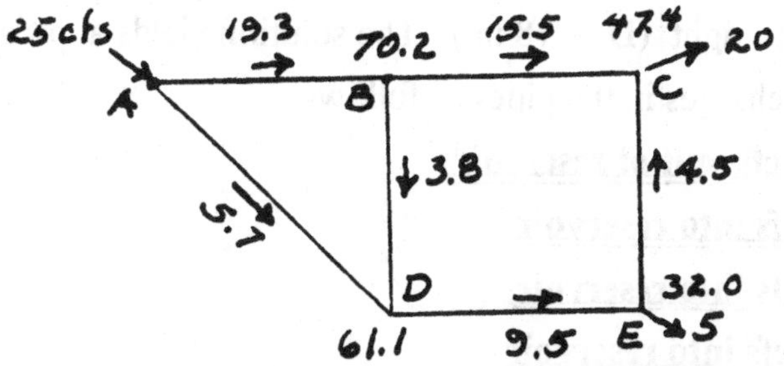

5-33. Continuity Eq. $Q_{AB} + Q_{CB} = Q_{BD}$

Head loss equations:

$h_{f,AB} = (0.02)\ ((10{,}000)/(10/12))\ (Q_{AB}{}^2/(2gA^2);\quad A_{AB} = (\pi/4)\ (10/12)^2$

$= 0.545\ ft^2$

$h_{f,AB} = 12.53\ Q_{AB}{}^2$

or $Q_{AB} = 0.283\ \sqrt{h_{f,AB}}$

$h_{f,BD} = (0.02)\ (5{,}000/1)\ (Q_{BD}{}^2/2gA^2)) = 2.52\ Q_{BD}{}^2$

So $Q_{BD} = 0.630\ \sqrt{h_{f,BD}}$

$h_{f,CB} = 0.02\ (4{,}000/(8/12))\ (Q_{CB}{}^2/(2gA^2)) = 15.3\ Q_{CB}{}^2$

So $Q_{CB} = 0.256\ \sqrt{h_{f,CB}}$

Iterate to solve for Q:

h_B (assume) (ft)	$h_{f,AB}$ (ft)	Q_{AB} (cfs)	$h_{f,CB}$ (ft)	Q_{BC} (cfs)	h_{LBD} (ft)	Q_{BD} (cfs)	$Q_{AB}+Q_{CB}-Q_{BD}$ (cfs)
50.0	100.0	2.83	50.0	1.81	50.0	4.46	0.18
55.0	95.0	2.75	45.0	1.72	55.0	4.67	-0.20
52.5	97.5	2.79	47.5	1.76	52.5	4.57	-0.02 (Close enough)

5-34. Because the water surface in the reservoir on the left is higher than all others, it is obvious that flow must be discharging from it. Likewise, flow must be going in the reservoir on the right. Therefore, p/γ at the junction must be greater than 50 ft and less than 100 ft. Iterate on p/γ at the junction until the continuity equation is satisfied at the junction (inflow = outflow). Let the pipes be identified as A, B, C and D from left to right (D = 18 in.). The solution yields a p/γ at the junction of **83.7 ft**, and discharges in the pipes as follows:

Q_A = 11.2 cfs out of reservoir

Q_B = 1.6 cfs into reservoir

Q_C = 2.5 cfs into reservoir

Q_D = 7.1 cfs into reservoir

5-35. Head loss equations:

$h_{f,AB} = f(L/D)\ Q^2/(2gA^2)$ where $A = (\pi/4)\ (8/12)^2 = 0.349\ ft^2$

$k_s/D = 0.0006$ (Fig. 5-5)   Assume $f = 0.020$

$f_{f,AB} = 0.020\ (400/(8/12))\ Q^2/(2g \cdot 0.349^2) = 1.53\ Q_{AB}^2$

$Q_{AB} = 0.808\ \sqrt{h_{f,AB}}$ (1)

$h_{f,DB} = f\ (L/D)\ Q^2/(2gA^2)$ where $A_{DB} = (\pi/4)\ (10/12)^2 = 0.5454\ ft^2$

$k_s/D = 0.0005$ (Fig. 5-5)   Assume $f = 0.019$

Then $h_{f,DB} = 0.019\ (600/(10/12))\ Q^2/(2g \cdot 0.5454^2) = 0.714\ Q_{DB}^2$

$Q_{DB} = 1.183\ \sqrt{h_{f,DB}}$ (2)

$h_{f,BC} = f(L/D)\ Q^2\ /(2gA^2)$ where $A_{BC} = (\pi/4) \cdot 1^2 = 0.785\ ft^2$

$k_s/D = 0.0004$ Assume $f = 0.018$

$h_{f,BC} = 0.018\ (1{,}000/1)\ Q^2/(2g \cdot .785^2) = 0.454\ Q_{BC}^2$

$Q_{BC} = 1.485\ \sqrt{h_{f,BC}}$ (3)

The above equations neglect minor losses and the velocity head at C. Check to see if flow is into or out of lower reservoir by assuming the piezometric head at B = 100 ft. above datum. Then by inspection of Eqs. (2) and (3) one can see that there will be more discharge in pipe BC than in pipe DB. Therefore, the piezometric head at B will be less than 100 ft. to yield a solution and the flow will be out of the lower reservoir.

Continuity equation: $Q_{AB} + Q_{DB} = Q_{BC}$

Iterate to obtain solution

h_B (ft)	$h_{f,AB}$ (ft)	Q_{AB} (cfs)	$h_{f,DB}$ (ft)	Q_{DB} (cfs)	$h_{f,BC}$ (ft)	Q_{BC} (cfs)	$Q_A+Q_B-Q_C$ (cfs)	
95.0	5.0	1.81	55.0	8.77	45.0	9.96	+0.62	
97.0	3.0	1.40	53.0	8.61	47.0	10.18	-0.17	
96.5	3.5	1.51	53.5	8.65	46.5	10.13	+0.03	<-- close enough

A more precise solution could be obtained by getting more precise values for f's and by considering the velocity head at C.

===

5-36. $Q = Q_{14} + Q_{12} + Q_{16}$

$20 = V_{14} \times (\pi/4) \times (14/12)^2 + V_{12} \times (\pi/4) \times 1^2 + V_{16} \times (\pi/4) \times (16/12)^2$; (1)

Also, $h_{f_{14}} = h_{f_{12}} = h_{f_{16}}$ and assuming f = 0.03 for all pipes

$(3{,}000/14)\ V_{14}^2 = (2{,}000/12)\ V_{12}^2 = (3{,}000/16)\ V_{16}^2$ (2)

$V_{14}^2 = 0.778\ V_{12}^2 = 0.875\ V_{16}^2$

From Eq(1) $20 = 1.069\ V_{14} + 0.890\ V_{14} + 1.49\ V_{14}$; $V_{14} = 5.79$ ft/s

and $V_{12} = 1.134\ V_{14} = 6.56$ ft/s; $V_{16} = 6.19$ ft/s

$\underline{Q_{12} = 5.15\ \text{ft}^3/\text{sec}}$; $\underline{Q_{14} = 6.19\ \text{ft}^3/\text{s}}$; $\underline{Q_{16} = 8.64\ \text{ft}^3/\text{s}}$

$V_{29} = Q/A_{29} = 20/(\pi/4 \times 1.21^2) = 4.36$ ft/s; $V_{30} = 4.074$ ft/s

$h_{L_{AB}} = (0.03/64.4)[(2{,}000/1.21)(4.36)^2 + (2{,}000/1) \times (6.56)^2$

$+ (4{,}000/(30/12) \times (4.074)^2] = \underline{60.0\ \text{ft}}$

5-37. First, write the energy equation from the center of the main pipe to the water surface of the river.

$p_p/\gamma + V_p^2/2g + z_p = p_{W.S.}/\gamma + V_{W.S.}^2/2g + \Sigma h_L$

$p_p/\gamma + 0 + 490 = 0 + 0 + 500 + V_{r.p.}^2/2g\ (0.5 + 1.0 + fL/D)$ (1)

where $V_{r.p.}$ = velocity in riser pipe = Q/A = 22.91 ft/sec

$Re_{r.p.} = (VD/\nu)_{r.p.} = 22.91 \times (1/3)/(0.74 \times 10^{-5})$

$= 1.0 \times 10^6$

Note: the kinematic viscosity was based upon an assumed effluent water temperature of 100°F.

$k_s/D = 0.0004$ (Fig. 5-5, assumed steel pipe)

$f_{r.p.} = 0.017$ (Fig. 5-4)

Then the head loss in the end 4-inch riser pipe is given as

$h_{L,r.p.} = (22.91^2/6.44)\ (1.5 + (0.017 \times 2(1/3)))$ (2)

$= 13.07$ ft.

5-37.(continued).

Then from Eq. (1) $p_p/\gamma = 500 - 490 + 13.07 = 23.07$ ft.

Also from Eq. (1) and Eq. (2)

$$h_p - 10 = (Q^2/(2gA^2))\,(1.60) \qquad \text{but } A = 0.872 \text{ ft.}^2$$

So $h_p - 10 = 3.27\,Q^2$

or $Q_{r.p.} = 0.553\sqrt{h_p - 10}$ (3)

Now consider the head loss in the main pipe between the last two riser pipes.

$Q = 2$ cfs so $V = Q/A = 2.55$ ft/sec.

$Re = VD/\nu = 2.55 \times 1/(.74 \times 10^{-5}) = 3.5 \times 10^5$

$k_s/D = 0.00016$; $f = 0.016$ (Fig. 5-4)

Then $h_f = 0.016\,(10/1) \times 2.55^2/64.4 = 0.02$ ft.

The remainder of the calculations are shown in tabular form (below) to yield the head in the main pipe at riser number 10.

i Junction pt.	h_p (ft)	h_f (ft)	Q riser (cfs)	Q pipe (cfs)	V pipe (ft)	f
1	23.07		2.00			
		0.020		2.00	2.55	.0160
2	23.09		2.00			
		0.060		4.00	5.10	.0150
3	23.15		2.01			
		0.127		6.01	7.65	.0140
4	23.27		2.01			
		0.227		8.02	10.22	.0140
5	23.50		2.03			
		0.356		10.05	12.80	.0140
6	23.86		2.06			
		0.517		12.11	15.42	.0140
7	24.38		2.10			
		0.712		14.21	18.10	.0140
8	25.09		2.15			
		0.907		16.36	20.80	.0135
9	26.00		2.21			
		1.173		18.57	23.66	.0135
10	27.17		2.29			
				20.86	26.57	.0135

5-37. (continued)

Now determine the water surface elevation in the reservoir.

$p_r/\gamma + V^2/2g + z_r = p_p/\gamma + V_p^2/2g + z_p + \Sigma h_L$

$0 + 0 + z_r = 27.17 + 26.57^2/64.4 + 490 + (26.57^2/64.4)(0.5 + fL/D)$

$z_r = 27.17 + (26.57^2/64.4)(1.5 + .0135 \times 500)$

$z_r = \underline{607.6}$ ft.

5-39. $r = 8 - y$

y(in.)	r(in.)	V(ft/s)	$2\pi rV(ft^2/s)$	area(ft^3/s)
0.0	8.0	0	0	
0.1	7.9	7.2	29.78	.124
0.2	7.8	7.9	32.26	.258
0.4	7.6	8.8	35.02	.561
0.6	7.4	9.3	36.03	.592
1.0	7.0	10.0	36.65	1.211
1.5	6.5	10.6	36.08	1.515
2.0	6.0	11.0	34.56	1.472
3.0	5.0	11.7	30.63	2.716
4.0	4.0	12.2	25.55	2.341
5.0	3.0	12.6	19.79	1.889
6.0	2.0	12.9	13.51	1.388
7.0	1.0	13.2	6.91	.851
8.0	0.0	13.5	0	.288

$Q = 15.21\ ft^3/s$

$V_{mean} = Q/A = 15.21/(0.785(1.33)^2) = 10.9$ ft/s

$V_{max}/V_{mean} = 13.5/10.9 = 1.24$; flow is turbulent because $V_m/V_{mean} < 2.0$

5-40. $\Delta h = 9$ m of water; $Q = 2 m^3/s$; D = 1 m; K = 0.65 (assume)

Then $d^2 = (4/\pi) \times 2/((0.65\sqrt{2g \times 9})) = 0.295$; d = 0.54 m

Try again: d/D = 0.54; $Re = 4Q/(\pi d\nu) \simeq 4.7 \times 10^6$; K = 0.63 (Fig. 5-21)

Then $d = \sqrt{0.65/0.63} \times 0.54 = 0.548$ m; K = 0.63 (same)

Thus $\underline{d = 54.8 \text{ cm}}$

5-41. Assume K = 0.65; T = 20°C; $\Delta h = \Delta p/\gamma$ = 50,000/9,790 = 5.11 m

Then following the procedure for P5-40:

$d^2 = (4/\pi) \times 3.0/(0.65\sqrt{2 \times 9.81 \times 5.11}) = 0.587$; d = 0.766 m

Check K: $Re_d = 4Q/(\pi d\nu) = 4 \times 3.0/(\pi \times 0.766 \times 10^{-6}) = 5 \times 10^6$

d/D = 0.766/1.2 = 0.64 Thus, K = 0.67 (from Fig. 5-21)

Try again: $d = \sqrt{(0.65/0.64)} \times 0.766 = 0.778$

Check K: $Re_d = 5 \times 10^6$ and d/D = 0.65 so K = 0.675 (Fig. 5-21)

$d = \sqrt{(0.67/0.675)} \times 0.778 = \underline{0.775 \text{ m}}$

5-42. $\Delta h = \Delta p/\gamma$ = 10 lbs/in^2 (144 lb/ft^2/16/in^2) /62.4 lb/ft^3 = 23.08 ft.

d/D = 0.50; Assume K = 1.02 (Fig. 5-21)

Then $Q = KA\sqrt{2g\Delta h}$

$= 1.02\ ((\pi/4)\ (2^2))\ \sqrt{(64.4)\ (23.08)}$

= 123.5 cfs

Check K: $Re_d = 4Q/(\pi d\nu) = 4 \times 123.5/\ (\pi \times 2 \times 1.2 \times 10^{-5}) = 6.5 \times 10^6$

Then from Fig. 5-21 it is seen that K = 1.02, so solution is correct.

<u>Q = 123.5 cfs</u>

5-43. d/D = 0.60

$Re_d = 4Q/(\pi d\nu) = 4 \times 3/(\pi \times 0.5 \times 1.22 \times 10^{-5}) = 6.3 \times 10^5$

from Fig. 5-21: K = 0.65; $A = (\pi/4) \times 0.5^2 = 0.196$ ft^2

Then $\Delta h = (Q/KA)^2/2g = (3/(0.65 \times 0.196))^2/64.4$ = 8.61 ft of water

$h = \Delta h/12.6$ = 0.683 ft = <u>8.2 in.</u>

5-44. Assume K = 1.01; Assume T = 20°C

$Q = KA\sqrt{2g\Delta h}$ where Δh = 200,000 Pa/9,790 N/m^3 = 20.4 m

Then $A = Q/(K\sqrt{2g\Delta h})$ or $\pi d^2/4 = Q/(K\sqrt{2g\Delta h})$

$d = (4Q/(\pi K\sqrt{2g\Delta h}))^{1/2}$

$d = (4 \times 10/(\pi \times 1.01\sqrt{2g \times 20.4}))^{1/2} = 0.794$ m

Check K: $Re = 4Q/(\pi d\nu) = 1.6 \times 10^7$; $d/D = 0.4$ so $K \approx 1.0$ (from Fig. 5-21)

Try again: $d = (1.01/1.0)^{1/2} \times 0.794$ = 0.798 m

5-45. $Re_d = 4 \times 0.57/(\pi \times 0.30 \times 1.49 \times 10^{-5}) = 1.6 \times 10^5$; $d/D = 0.50$; $K = 1.00$

$\Delta h = (Q/(KA))^2/(2g) = (0.57/(1 \times (\pi/4) \times 0.3^2)^2/(2 \times 9.81) = 3.32$ m

deflection h = 3.32/12.6 = 0.263 m

5-46. $\sigma = E\Delta T\alpha$ (Eq. 5-38)

$\sigma = (4.5 \times 10^6$ psi) (80°F - 40°F) (6×10^{-6} ft/ft/°F)

σ = 1080 psi

5-47. $S_{safe} = (S_{3\text{-}edge} \times F_{load})/F_{safety}$ (5-41)

Assume $F_{safety} = 2.0$

$F_{load} = 1.5$ (from Fig. 5-27)

Thus $S_{safe} = (2.0/1.5)\, S_{3\text{-}edge} = 1.33\, S_{3\text{-}edge}$

Applied load: $W = C\gamma_s B_d^2$ (Eq. 5-39 for rigid pipe)

H = 10 ft - 2 ft = 8 ft so $H/B_D = 8/3 = 2.67$

C = 2.0 (Fig. 5-28 for $H/B_d = 2.67$, clay)

Assume γ_s = 150 lbs/ft^3

$W = 2.0 \times 150 \times 3^2$ = 2700 lbs/ft.

S_{safe} must be greater than W = 2700 lbs/ft

or $S_{3\text{-}edge} = W/1.33$ (from Eq. (1)) or $S_{3\text{-}edge} = 2700/1.33$ = 2,030 lb/ft

Use Class I pipe (from Table 5-5).

5-48. $V = Q/A = 31.4/(\pi \times 1 \times 1) = 9.995$ ft/sec

$$\Sigma F_y = \rho Q(V_{2y} - V_{1y})$$

$$F_{anch} - Wt_{water} - Wt_{bend} - p_2A_2 \sin 30° = \rho Q(V \sin 30° - V \sin 0°)$$

$$F_{anch} = \pi \times 1 \times 1 \times 4 \times 62.4 + 300 + 8 \times 144 \times \pi \times 1 \times 1 \times 0.5$$

$$+ 1.94 \times 31.4 \times (9.995 \times 0.5 - 0)$$

$$= \underline{3,198 \text{ lbs}}$$

5-49. $(V_1^2/2g) + (p_1/\gamma) + z_1 = (V_2^2/2g) + (p_2/\gamma) + z_2$

$(V_1^2/2g) + (25 \times 144/62.4) + 0 = 16(V_1^2/2g) + 0 + 0$

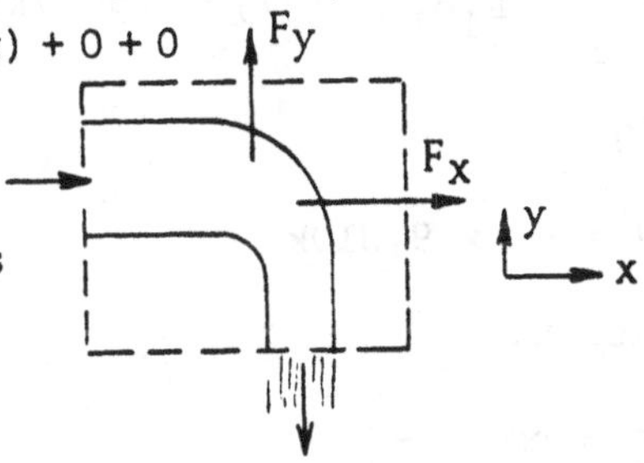

$V_1 = 15.74$ fps; $V_2 = 62.95$ ft/sec

$Q = A_1V_1 = \pi \times 1 \times 1 \times 15.74 = 49.44$ cfs

$\Sigma F_x = \rho Q(V_{2x} - V_{1x})$

$p_1A_1 + F_x = \rho Q(V_{2x} - V_{1x})$

$F_x = 1.94 \times 49.44(0 - 15.74) - 25 \times 144 \times \pi \times 1 \times 1 = \underline{\underline{-12,820 \text{ lbf}}}$

5-50. $V = Q/A = 0.4/(\pi \times 0.2 \times 0.2) = 3.183$ m/s

$F_x = -90,000 \times \pi \times 0.2 \times 0.2(1 + \cos 45°) + 1,000 \times 0.4$

$\times 3.183(-\cos 45° - 1) = -21,480 \text{ N} = \underline{\underline{-21.480 \text{ kN}}}$

$F_y = -90,000 \times \pi \times 0.2 \times 0.2 \sin 45° + 1,000 \times 0.4(-3.183 \sin 45°)$

$= -8,897 \text{ N} = \underline{-8.897 \text{ kN}}$

5-51.

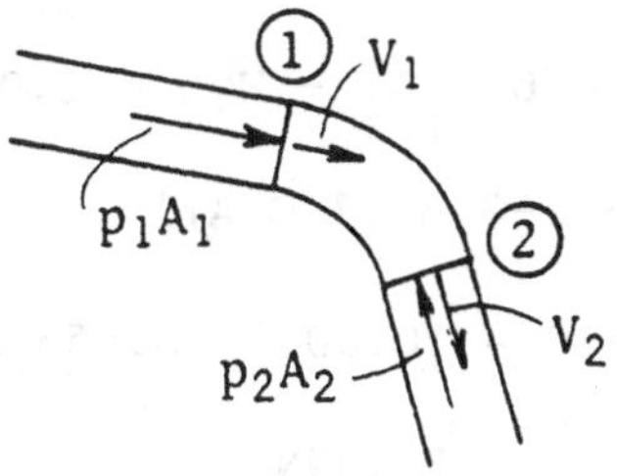

5-51. (continued).

Preliminaries:

Velocities:

$\underline{V}_1 = (Q/A)[(13/\ell_1)\underline{j} - (10/\ell_1)\underline{k}]$ where $\ell_1 = \sqrt{13^2+10^2}$

Thus $\underline{V}_1 = (Q/A)[0.793\,\underline{j} - 0.6097\,\underline{k}]$

$\underline{V}_2 = (Q/A)[(13/\ell_2)\underline{i} + (19/\ell_2)\underline{j} - (20/\ell_2)\underline{k}]$ where $\ell_2 = \sqrt{13^2+19^2+20^2}$

Then $\underline{V}_2 = (Q/A)[0.426\underline{i} + 0.623\underline{j} - 0.656\underline{k}]$

Pressure forces:

$\underline{F}_{p_1} = p_1A_1(0.793\underline{j} - 0.6097\underline{k})$; $\underline{F}_{p_2} = p_2A_2(-0.426\underline{i} - 0.623\underline{j} + 0.656\underline{k})$

Weight:

$\underline{W} = -3 \times 9{,}810\underline{k}$

Solution:

$\Sigma\underline{F} = \rho Q(\underline{V}_2 - \underline{V}_1)$

$F_{block,x} - 0.426\, p_2A_2 = \rho Q[0.426\, Q/A - 0]$; where $p_2A_2 = 25{,}000$

$\times (\pi/4) \times (1.3)^2 = 33{,}183$ N; $Q/A = 15/((\pi/4) \times (1.3)^2) = 11.30$ m/s

Then $F_{block,x} = 1{,}000 \times 15 \times 0.426 \times 11.3 + 0.426 \times 33{,}183 = \underline{\underline{86{,}343\text{N}}}$

$F_{block,y} + 0.793\, p_1A_1 - 0.623\, p_2A_2 = \rho Q[0.623(Q/A) - 0.793\, Q/A]$

where $p_1A_1 = 20{,}000 \times (\pi/4)(1.3)^2 = 26{,}546$ N

Then $F_{block,y} = 1{,}000 \times 15(11.3)(-0.170) - 0.793 \times 26{,}546$

$+ 0.623 \times 33{,}183 = -28{,}815 - 21{,}051 + 20{,}673 = \underline{\underline{-29{,}193\text{N}}}$

$F_{block,z} - 0.6097\, p_1A_1 + 0.656\, p_2A_2 - Wgt = 1{,}000$

$\times 15[-0.656(Q/A) - (-0.6097\, Q/A)]$

$F_{block,z} = -7{,}848 + 3 \times 9{,}810 + 10{,}000 - 0.656 \times 33{,}183$

$+ 0.6097 \times 26{,}546 = \underline{\underline{25{,}999\text{N}}}$

Then the total force which the thrust block exerts on the bend to hold it in place is

$\underline{\underline{\underline{F} = (86.35\underline{i} - 29.19\underline{j} + 26.00\underline{k})\text{kN}}}$

5-52. Take section 1 at reservoir surface and section 2 at section of d diameter.

$p_1/\gamma + V_1^2/2g + z_1 = p_2/\gamma + V_2^2/2g + z_2$

$0 + 0 + 5 = p_{2,vapor}/\gamma + V_2^2/2g + 0$ where $p_{2,vapor} = 2{,}340$ Pa abs.

$p_{2,vapor} = -97{,}660$ Pa gage

Then $V_2^2/2g = 5 + 97{,}660/9{,}790 = 14.97$ m; $V_2 = 17.1$ m/s

$Q = V_2A_2 = 17.1 \times \pi/4 \times 0.15^2 = \underline{0.303\ m^3/s}$

5-53. First write energy equation from water surface in reservoir to outlet of sluices:

$p_1/\gamma + V_1^2/2g + z_1 = p_2/\gamma + V_2^2/2g + z_2 + \Sigma h_L$

$0 + 0 + 100 = 0 + V_6^2/2g + 30 + f\ (L/D)\ V_8^2/2g$

Assume $f = 0.015$; $A_6 = (\pi/4)\ 6^2 = 28.2\ ft^2$; $A_8 = 50\ ft^2$

$V_6^2/2g = Q^2/((28.2^2)\ (64.4)) = 1.95 \times 10^{-5}\ Q^2$; $V_8^2/2g = 0.62 \times 10^{-5}\ Q^2$

Then from Eq. (2): $70 = Q^2 \times 10^{-5}\ (1.95 + (0.62)\ (0.015 \times 200/8))$

$Q = 1790$ cfs $\quad V_6 = Q/A_6 = 63.5$ ft/sec

Check f: $Re = VD/\nu = 63.5 \times 6/\ (1.2 \times 10^{-5}) = 3.2 \times 10^7$

$k_s/D \approx 0.00002$ (Fig. 5-5); $f = 0.009$ (Fig. 5-4)

Recalculate with $f = 0.009$:

$70 = Q^2 \times 10^{-5}\ (1.95 + (0.62)\ (0.009 \times 200/8))$

$Q = \underline{1830\ cfs} \quad V_6 = 64.9$ ft/sec $\quad V_8 = 36.6$ ft/sec

Now consider the force at the joint

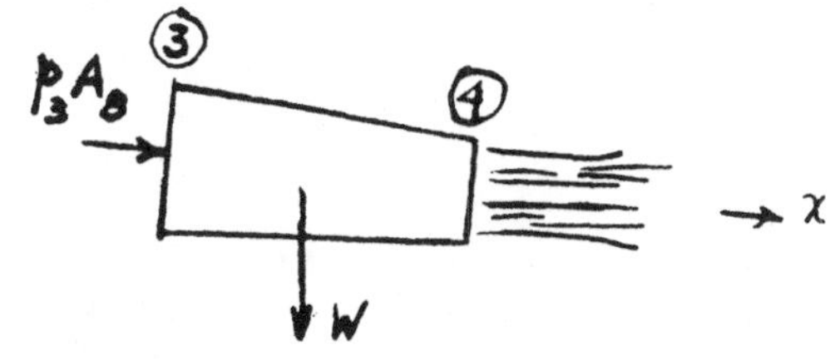

$\Sigma F_x = \rho Q\ (V_6 - V_8)$

$p_3A_8 + F_{joint} = 1.94 \times 1830\ (64.9 - 36.3)$

$F_{joint} = -p_3A_8 + 101{,}536$

Get p_3: $p_3/\gamma + V_8^2/2g \approx V_6^2/2g$

5-53. (continued).

$p_3 = \gamma(V_6^2/2g - V_8^2/2g)$

$= 62.4\ ((64.9^2/2g) - (36.6^2/2g)) = \underline{2783\ psf}$

Thus $F_{joint} = -2783 \times 50 + 101{,}536 = \underline{-37{,}614\ lbs}$

In addition a force will have to be applied at the joint to resist the weight of the nozzle and weight of water in the nozzle. Depending upon the length of the nozzle this may be as much as 30% or 40% of the force calculated above and it will act upward.

5-54. $V_{18} = Q/A_{18} = 25/((\pi/4) \times 1.5^2) = 14.15$ ft/s

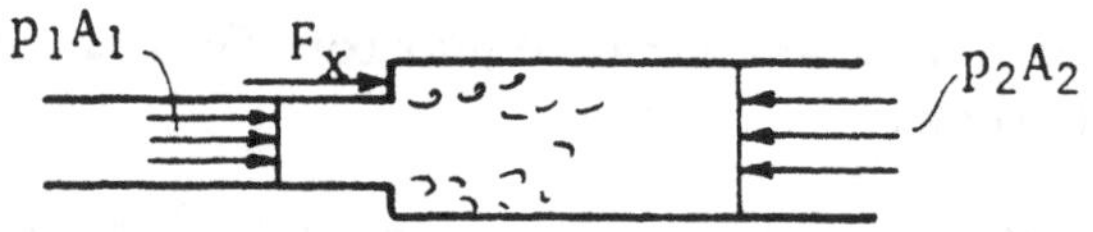

$V_{24} = Q/A_{24} = 7.958$ ft/s

$h_L = (14.15 - 7.958)^2/64.4 = \underline{0.595\ ft}$

$p_{18}/\gamma + V_{18}^2/2g = p_{24}/\gamma + V_{24}^2/2g + h_L$

Then $p_{24}/\gamma = p_{18}/\gamma + V_{18}^2/2g - V_{24}^2/2g - h_L$

$= 10 \times 144/62.4 + (14.15)^2/64.4 - (7.958)^2/64.4 - 0.595$

$p_{24}/\gamma = 24.61$ ft; $p_{24} = 24.61 \times 62.4 = 1{,}535\ lb/ft^2$

Now write the momentum equation for the transition

$\Sigma F_x = \rho Q(V_{24_x} - V_{18_x})$

$p_{18}A_{18} - p_{24}A_{24} + F_x = \rho Q(V_{24_x} - V_{18_x})$

$F_x = -10 \times 144 \times \pi/4 \times 1.5^2 + 1{,}535 \times \pi/4 \times 2^2 + 1.94 \times 25(7.958 - 14.15)$

$= \underline{1{,}978\ lbs}$

5-55. $h_f = f(L/4R)\, V^2/2g$ Assume granular or brushed surface in fairly good condition; thus,

$k_s = 0.001$ ft (from Table 5-6).

$Q = 1000$ cfs; $V = Q/A = 1{,}000/(0.5\,\pi\, r^2 + 80)$

$= 1{,}000/(0.5\,(25\pi) + 80)$

$= 1{,}000/(119.3) = 8.384$ ft/sec

$R = A/P = 119.3/(26 + 5\pi) = 2.86$ ft.

$Re = V(4R)/\nu = 8.384\,(4 \times 2.86)/(1.2 \times 10^{-5}) = 8 \times 10^6$

$k_s/4R = 0.001/(4 \times 2.87) = 8.74 \times 10^{-5}$

$f \approx 0.012$ (Fig. 5-4)

Then $h_f/1{,}000$ ft $= 0.012\,(1000/(4.\times 2.87)\,(8.384^2/64.4))$

$=$ 1.14 ft

5-56. Tunnel dimensions:

$A = (0.5\,\pi\, r^2 + 180)$

$A = (0.5\,\pi\,(7.5^2) + 180) = 268.4$ ft^2

$P = (\pi r + 24 + 15) = (\pi\,(7.5) + 39) = 62.56$ ft.

$R = A/P = 4.29$ ft.

Assume $k_s = 1.4$ ft; $k_s/4R = 1.4\,/(4 \times 4.29) = 0.17$

$f \approx 0.048$ (Fig. 5-4)

$h_f = f\,(L/4R)\,V^2/2g$

$2 = 0.048\,(1000/(4 \times 4.29))\,Q^2/(2gA^2)$

$Q =$ 1820 cfs

6-1. Consider the resultant force R as shown. Resolve this force into components Ry and Rx acting on the base. Let x be the distance from the centroid of the base to where the resultant force intersects the base. There the flexure formula will yield the normal stress at any point on the base:

$\sigma = P/A \pm Mc/I$

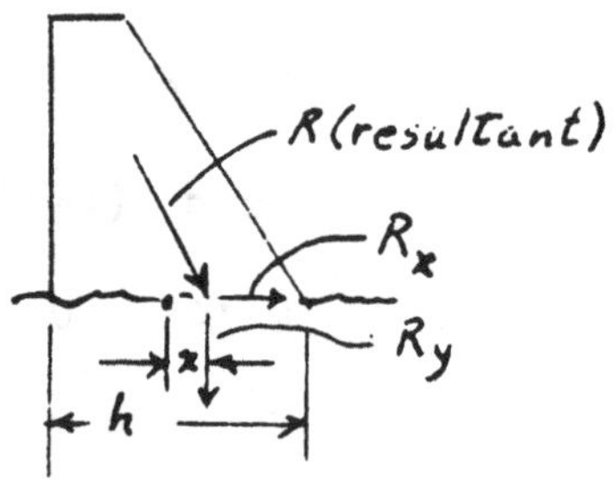

where $P = R_y$

$A = h$ (considering unit length of dam)

$M = R_y x$

$I = 1 \times h^3/12$ (for unit length of dam)

$\sigma = R_y/h \pm R_y x\ (h/2)\ /\ (h^3/12)$

However, for the condition of zero tensile stress $\sigma = 0$; therefore,

$0 = R_y/h \pm R_y x(6/h^2)$

solving for x yields $x = \pm\ h/6$

Thus, for $x \pm h/6$ (limits of middle third of the base) the n stress will be zero at the heal or toe but for any other po within the middle third the normal stress will be positive.

6-2. Assume concrete = 150 lb/ft^3

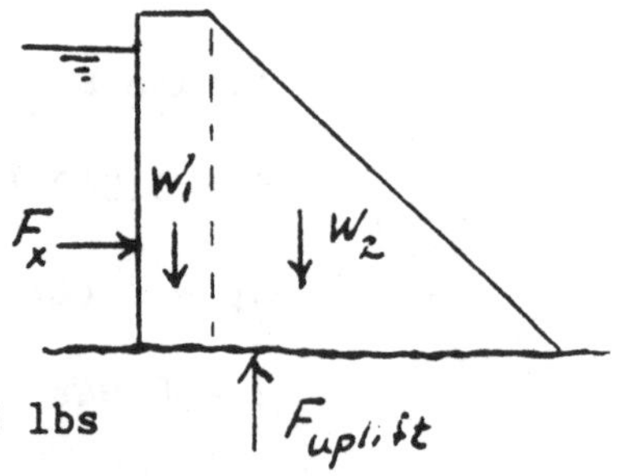

Compute forces per unit length of dam:

$W_1 = \gamma V_1 = 150\ (10 \times 50) = 75{,}000$ lbs

$W_2 = \gamma V_2 = 150\ ((.5) \times 50 \times 35) = 131{,}250$ lbs

$F_x = pA = 62.4 \times (45/2) \times 45 = 63{,}180$ lbs

$(2/3)\ F_{uplift} = (2/3)\ ((1/2)\ (62.4)\ (45)\ (45)) = 42{,}120$ lbs

$\sigma_y = \Sigma P/A \pm \Sigma Mc/I$

$\Sigma P = W_1 + W_2 - (2/3)\ F_{uplift}$

$= 75{,}000 + 131{,}250 - 42{,}120 = 164{,}130$ lbs

$A = 45$ ft^2

6-2 (continued).

$$\Sigma M = -15\ F_x + 17.5\ W_1 + 0.833W_2 - 7.5\ (2/3\ F_{uplift})$$

$$= -15\ (63{,}180) + 17.5\ (75{,}000) + 0.833\ (131{,}250) - 7.5\ (42{,}120)$$

$$= 947{,}700 + 1{,}312{,}500 + 109{,}375 - 315{,}900$$

$$= 158{,}275 \text{ ft lbs.}$$

Consider normal stress at heel:

$$\sigma_{heel} = 164{,}130/45 - 158{,}275 \times (45/2)\ /\ (45^3/12)$$

$$= 3647 + 469 = \underline{4{,}116} \text{ psf (in compression)}$$

$$\sigma_{toe} = 3647 - 469 = \underline{3{,}178} \text{ psf (in compression)}$$

6-3 Assume earthquake acceleration of 0.1 g. Assuming acceleration direction toward reservoir there will be an added clockwise moment due to the acceleration of the mass of the dam and there will be an added force due to the additional pressure of the water on the dam. Assume no change in uplift force.

Added moment from mass of dam:

W_1: $F = ma$

$$= (W_1/g)\ (0.1\ g) = 0.1\ W_1$$

moment arm = 25 ft
added moment = $-25 \times 0.1\ W_1 = -187{,}500$ ft-lb

W_2: $F = 0.1\ W_2$

added moment = $-0.1\ W_2\ (50/3) = -218{,}750$ ft-lb

Added moment from added hydrostatic force:

$$M_e = 0.299\ P_e y^2 \qquad (6\text{-}4)$$

where $P_e = C\lambda\gamma h$

$$C = 0.365\ [(y/h)\ (2-(y/h)) + \sqrt{(y/h)\ (2-(y/h))}] \qquad (6\text{-}2)$$

$y = h = 45$ ft; $C = 0.73$

$\gamma = 62.4$ lb/ft^3; $\lambda = a/g = 0.1$

so $P_e = 0.73 \times 0.1 \times 62.4$ lb/ft$^3 \times 45$ ft $= 205$ lb/ft^2

Then $M_e = -0.299 \times 205 \times 45^2 = -124,\ 122$ ft-lb

6-3 (continued).

Total moment (including earthquake forces):

ΣM = +158,275 - 187,500 - 218,750 - 124,122

= -372,097 ft-lb

σ_{heel} = P/A + Mc/I

= 3,647 - 372,097 (45/2) / (45^3/12)

= 3,647 - 1103 = 2544 psf

σ_{toe} = 3,647 + 1103 = 4,750 psf

6-4. Assume f = 0.80 (Table 6-3)

$F_V = W_1 + W_2 - F_{uplift}$

= 75,000 + 131,250 - 42,120

= 164,130 lbs

F_x = 63,180 lbs

fF_V/F_x = 0.80 (164,130) / 63,180 = 2.08

since $fF_V/F_x > 1.0$ it is stable against sliding

6-5. t = h (L/2) / (cos (90° - (133.3/2)) (σ)) (6-7)

where L = 1,970 ft

h = 270 ft

= 62.4 lb/ft+2

= 2,000 psi = 288,000 psf

Thus t = 62.4 x 270 (1970/2) / (cos (90° - (133.3/2)) (288,000))

= 62.8 ft

6-6. Convert evaporation amounts to equivalent discharges in second-foot-months (sfm).

The volume of 1 sfm = 1 ft^3/s (30 da/mo) (24 hr/da) (3600 s/hr)

$$1 sfm = 2,592,000\ ft^3$$

$$= 59,504 \text{ acre ft}$$

x inches of evaporation from reservoir = 475 acres $(x/12)$ ft

$$= 39.58\ x \text{ A.F.}$$

Then x inches of evap. per mo. = 39.58 x A.F.

$$= 0.665\ x \text{ sfm}$$

Assume pan evap. coefficient = 0.70

Then Lake evap. = 0.70 x pan evap.

The table below gives lake evap. in inches (Col. 2), equivalent lake evap. in sfm (Col. 3) and creek inflow minus evaporation rate (Cols. 4-7).

Month	Lake evaporation (inches)	sfm	Streamflow corrected for evap. (cfs) 1975	1976	1977	1978
O	0.35	0.23	12.77	5.77	3.77	13.77
N	0.14	0.09	66.91	19.91	17.91	54.91
D	--	--	30.00	15.00	25.00	41.00
J	--	--	62.00	30.00	23.00	56.00
F	--	--	23.00	10.00	15.00	37.00
M	0.35	0.23	25.77	11.77	19.77	28.77
A	2.10	1.40	37.60	12.60	22.60	40.60
M	2.80	1.86	54.14	18.14	17.14	59.14
J	3.50	2.33	54.67	7.67	12.67	42.67
J	4.90	3.26	5.74	1.74	4.74	8.74
A	4.20	2.79	2.21	.21	3.21	4.21
S	2.80	1.86	5.14	.14	6.14	4.14

Next, plot the flows for the years of 1975, '76, '77. and '78 as a mass diagram as shown below. The storage available to augment the low flow is approx. 13,400 A.F. - 1,500 A.F. = 11,900 A.F. (From Fig. for prob.)

$$\frac{11,900 \text{ A.F.}}{59,504 \text{ A.F./sfm}} = 200 \text{ sfm}$$

Applying the storage ordinate of 200 sfm as shown in the figure(next pg.) yields an assured discharge of <u>20 cfs</u> for the drought period.

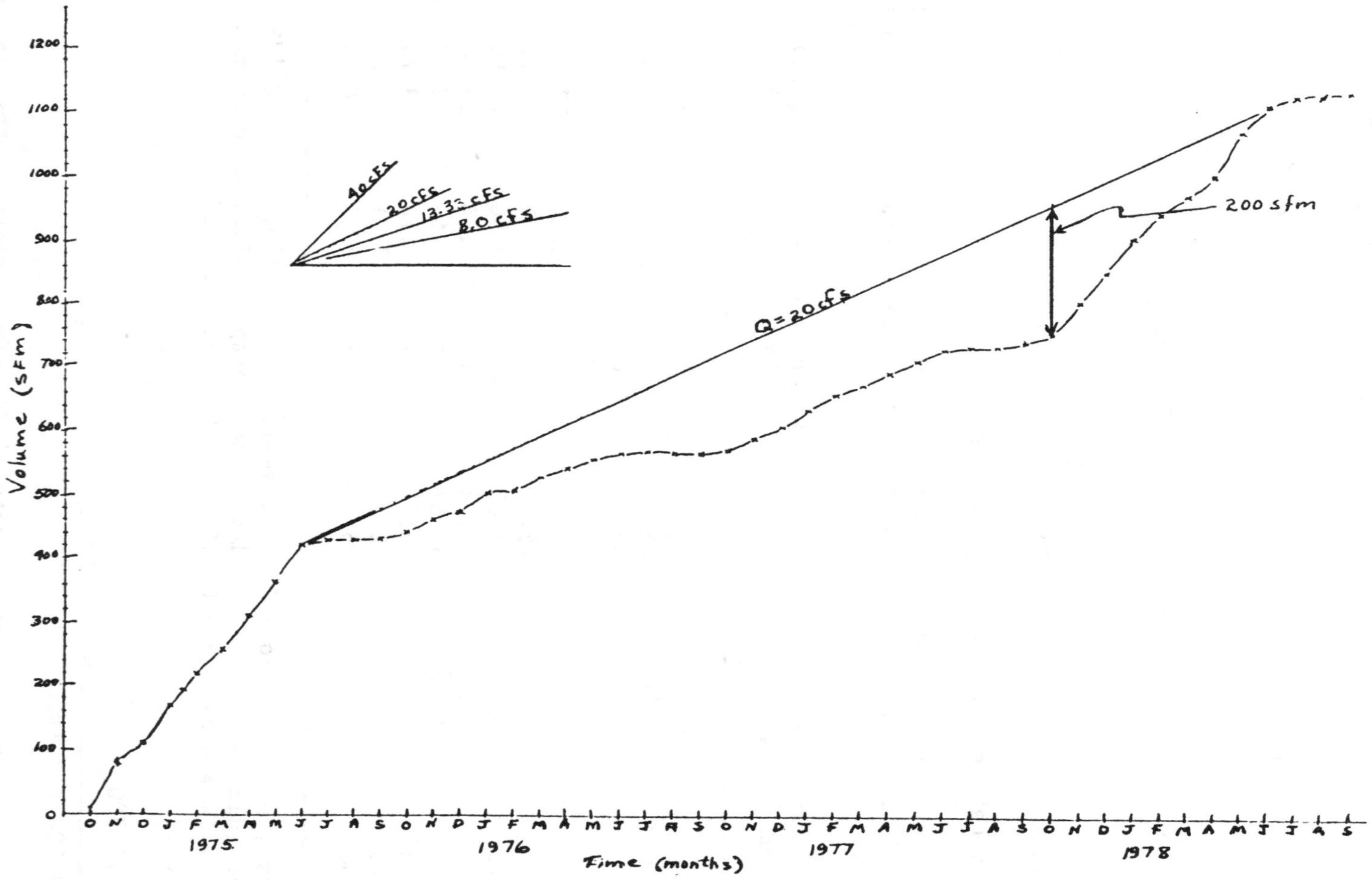

Fig. for Problem 6-6. Mass Diagram for the Wasco Reservoir (Oct. 1975-Oct. 1978)
(adapted from solution by Tim L. Martin)

6-7. Assume 70 mph wind = 102.7 ft/sec
Fetch = 9,000 ft (Fig. for P6-6) = 1.7 miles
Determine significant wave height.

Significant wave height = 4.0 ft (Fig. 6-31) and it takes approximately 22 minutes for these waves to develop. assume we are interested in the height of wave for which only 2% of the waves exceed it. Thus from Table 4-6 we find the ratio (H/H_s = 1.4) where H_s = significant wave height.

Thus $H = 1.4 \times H_s$
$= 1.4 \times 4.0 = \underline{5.60 \text{ ft}}$

Now determine runup:

Embankment slope = 1/2.5 = 0.40

$$L_o = 0.159\ gT^2$$

where $T = 0.429\ U_o^{0.44}\ F^{0.28}/g^{0.72}$

$= 0.429 \times 102.7^{0.44}\ 9{,}000^{0.28}/32.2^{0.72}$

$= 3.46$ s

Then $L_o = 0.159 \times 32.2 \times 3.46^2 = 61.3$ ft

$H/L_o = 5.60/61.3 = 0.091$

$R/H_o \approx 0.45$ (extrapolated from Fig. 6-31 assuming permeable rubble riprap)

$R = 0.45\ H_o = \underline{2.52 \text{ ft}}$

6-10. $Q_{AVg} = \Sigma\ Q/24$ hrs. The ΣQ and other calculations are shown in the table below:

Time	Q_{demand}	Q_{AVg}	$Q_{from\ stor}$	ΔV(gal)	ΣV(gal)
1A	800gpm	1425gpm	-625	-37,500	108,000
2	800		-625	-37,500	70,500
3	900		-525	-31,500	39,000
4	1,000		-425	-25,500	13,500
5	1,200		-225	-13,500	0
6	1,425		0	0	Storage Reserv. full
7	1,900		475	28,500	28,500
8	2,200		775	46,500	75,000
9	2,000		575	34,500	109,500
10	1,575		150	9,000	118,500
11	1,600		175	10,500	129,000
12	1,700		275	16,500	145,400
1P	1,500		75	4,500	150,000
2	1,300		-125	-7,500	142,500
3	1,400		-25	-1,500	141,000
4	1,600		175	10,500	151,500
5	1,800		375	22,500	174,000
6	2,300		875	52,500	226,500
7	1,800		375	22,500	249,000
8	1,500		75	4,500	253,500 max. vol.
9	1,200		-225	-13,500	240,000
10	1,000		-425	-25,500	214,500
11	900		-525	-31,500	183,000
12	800		-625	-37,500	145,500

$\Sigma Q = 34,200$

$Q_{Avg} = 34,200/24 = \underline{1425 \text{ gpm}}$

From column 6 (above) it is seen that the maximum volume required is 253,500 gallons.

6-11. $t = \gamma h\ (L/2) / (\cos (90° - \alpha)\ \sigma)$

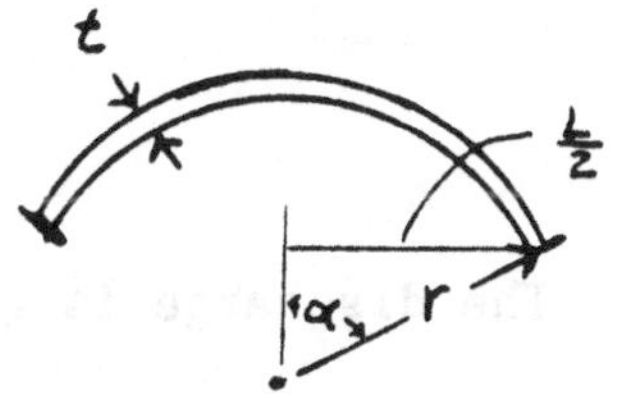

but $r\sin\alpha = L/2$

so $t = \gamma h\ r \sin\alpha/(\cos (90° - \alpha)\sigma)$

$\forall = tr\ (2\alpha)\Delta h$

$= \gamma h\ (r \sin \alpha)\ (r)\ (2\alpha) \Delta h/(\cos (90° - \alpha)\sigma)$

$= (\gamma h \Delta h/\sigma)\ r^2\ (2\alpha) \sin \alpha/ (\cos (90° - \alpha)\ \sigma)$

$\cos (90° - \alpha) = (L/2) /r$ (from Fig. above)

$\sin \alpha = (L/2)/r$

Thus $\forall = (\gamma h \Delta h/\sigma)\ 2r^2\alpha$

$\forall = K_1 r^2 \alpha$

$= K_1(L/2)^2\ \alpha/(\sin^2\alpha)$

$= K_2\ \alpha/\sin^2\alpha$

$= K_2\alpha\ \sin^{-2}\alpha$

$d\forall/d\alpha = 0 = K_2\ [- \alpha\ (2 \cos \alpha) / \sin^3\alpha + \sin^{-2}\alpha]$

$2\alpha / \tan \alpha = 1$

Solving for α yields $\alpha = \underline{66.784°}$

$2\alpha = \underline{133.57°}$

6-13. $S = 2.025 \times 10^{-6}\ V^2F/(gD)$ (6-12)

where $V = 40$ mph $= 58.67$ ft/s

$F = 37$ mi $= 195{,}360$ ft

$D = 7$ ft

Then $S = 2.025 \times 10^{-6} \times 58.67^2 \times 195{,}360/(32.2 \times 7)$

$= \underline{6.04\ \text{ft}}$

CHAPTER SEVEN

7-1. The discharge is given by Eq. (7-1): $Q = C\sqrt{2g}\ LH^{3/2}$

The head H = (1030 - 1000) + h_a, where h_a is the approach velocity head.

Figure 7-10(a) is used to determine C

For P/H = 200/30 = 6.67, C = C_D = 0.492

Neglecting h_a, $Q = C\sqrt{2g}\ LH^{3/2}$

$$150{,}000 = 0.492\ \sqrt{2g} \times L \times (1030-1000)^{3/2}$$

Solving, L = 231 ft

Assuming the channel width is also 231 ft,

V_a = 150,000/231 x 230 = 2.82 fps

$V_a^2/2g$ = 0.123

Including h_a, $Q = C\sqrt{2g}\ LH^{3/2}$

$$150{,}000 = 0.492\ \sqrt{2g} \times L \times (1030-1000+0.123)^{3/2}$$

Solving, L = 229 ft

Note: The effect of h_a is relatively minor in this application. The approach channel width is probably greater than the assumed 231 ft. If this is indeed the case, the crest length L would be slightly greater (perhaps 230 ft).

7-2. A spillway rating curve is the relationship between flow rate (Q) and water surface elevation. Figure 7-10(b) is used to develop the curve.

H/H_D	C/C_D	C	H (ft)	Q (cfs)	W.S. elevation (ft)
0			0	0	1000
0.1	0.82	0.403	3.0	3,850	1003
0.2	0.853	0.420	6.0	11,300	1006
0.3	0.879	0.432	9.0	21,400	1009
0.4	0.90	0.443	12.0	33,800	1012
0.5	0.92	0.453	15.1	48,900	1015
0.6	0.94	0.462	18.1	65,400	1018
0.7	0.956	0.470	21.1	83,700	1021
0.8	0.972	0.478	24.1	104,000	1024
0.9	0.988	0.486	27.1	126,000	1027
1.0	1.0	0.492	30.1	150,000	1030
1.1	1.012	0.498	33.1	174,000	1033
1.2	1.025	0.504	36.1	201,000	1036
1.3	1.038	0.511	39.1	230,000	1039
1.4	1.048	0.516	42.1	259,000	1042

7-3. The major dimensions for the spillway are determined from Figure 7-9. For H_D = 30' the results are:

H_D(ft)	n	R_1(ft)	R_2(ft)	x_C(ft)	y_C(ft)
30	1.872	15.9	7.05	8.49	3.81

7-4. Make the gates approximately rectangular in slope.
Then let W = 31 ft
H = 31 ft
Assume θ = 60°

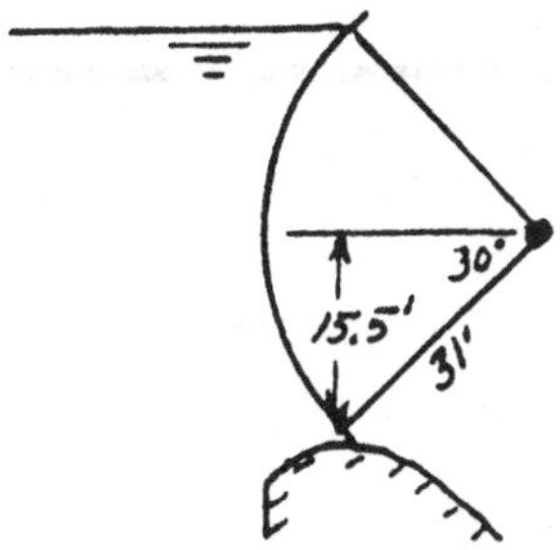

sin 30° = 15.5/r
r = 15.5/sin 30 = 31'

Choose No. of gates: 30 n = 231
n = 7.7 gates

Use 7 gates @ 33' wide

So make 7 gates 31 ft. high x 33 ft. wide

or could make 8 gates 31 ft. high x 29 ft. wide

Will have to have piers so make piers 3 ft. wide. So for 8 gates we need 7 piers

Then total width of spillway = 231 + 24 = 255 ft.

7-5. For a small opening, the velocity V immediately downstream from the gate has a horizontal component equal to $\sqrt{2gH}$ and zero vertical component.

The location of the jet downstream from the gate is determined from the parametric equations for a projectile as follows:

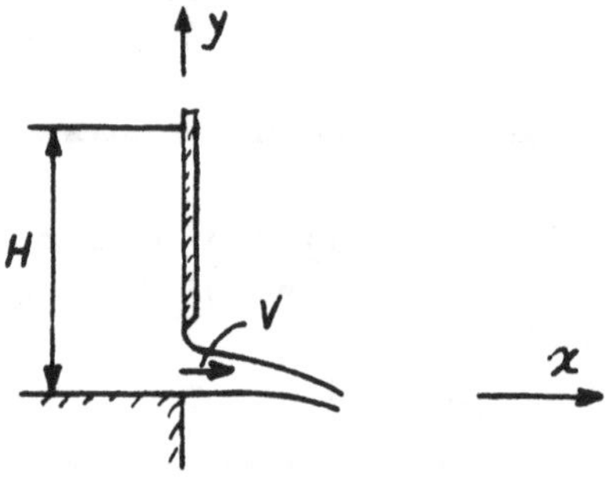

(1) The horizontal distance x is given by

$x = Vt = \sqrt{2gH}\ t$

(2) The vertical distance y is given by

$y = -gt^2/2$,

eliminating t, $y = -x^2/4H$

7-6. By careful design of the radial gate geometry (radius, location of gate sill, location of gate trunnion), a downward component may be imposed on the jet as it issues from the gate, thus furnishing a trajectory that more closely follows the shape of a spillway designed to correspond to the underside of a freely flowing weir flow. This will tend to raise pressures on the spillway face.

7-8. To find H_D, use Eq. 7-2: $Q = C\sqrt{2g}\,LH^{3/2}$

$200{,}000 = 0.492\,\sqrt{2g}\,(2000)\,H_D^{3/2} \dashrightarrow H_D = 8.62$ ft.

For $H/H_D = 0.25$, $C/C_D = 0.866$ (from Fig. 7-10(b)

Therefore,

$Q = (0.866)\,(0.492)\,\sqrt{2g}\,(2000)\,(0.25 \times 8.62)^{3/2}$

$= \underline{21{,}600 \text{ cfs}}$

For $H/H_D = 0.50$, $C/C_D = 0.92$

$Q = (0.92)\,(0.492)\,\sqrt{2g}\,(2000)\,(0.5 \times 8.62)^{3/2}$

$= \underline{65{,}000 \text{ cfs}}$

7-9. The radius $R = 0.85\,H_D = 0.85\,(40) = 34$ ft

$\cos\beta = 15/34$

$\beta = 63.8°$

From Fig. 7-23,

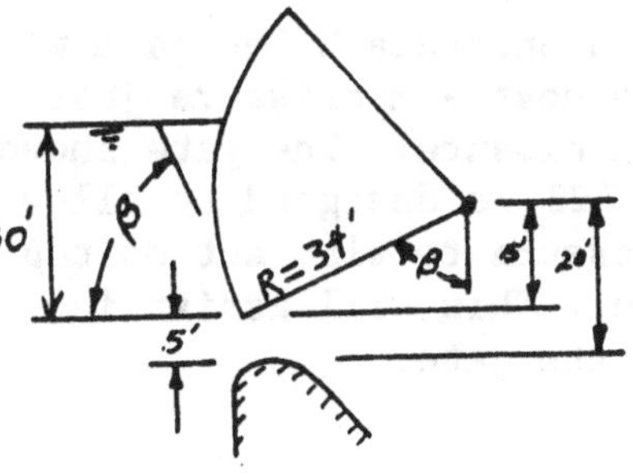

$\underline{C = 0.675}$

$d/H_1 = 5/30 = 0.167$

From Fig. 7-11

$\underline{C = 0.702}$

7-13. $V_o = \sqrt{2g\ (h_v + d)}$

$V_{o_x} = \cos\theta\sqrt{2g\ (h_v + d)}$

$V_{o_y} = \sin\theta\ \sqrt{2g\ (h_v + d)}$

$V_x = V_{o_x}$

$V_y = V_{o_y} - gt$

$x = V_{o_x}\ t = \cos\theta\sqrt{2g\ (h_v + d)}\ t$

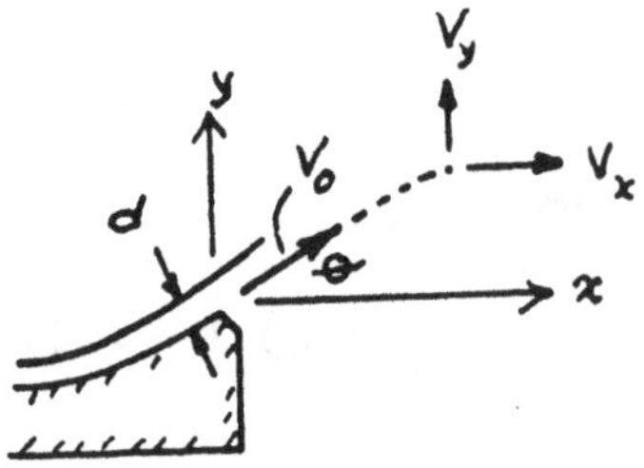

$$y = \int_0^t (V_{o_y} - gt)\ dt = V_{o_y}\ t - gt^2/2$$

$$= \sin\theta\ \sqrt{2g\ (h_v + d)}\ t - gt^2/2$$

Eliminating t,

$$y = x\tan\theta - x^2/4\cos^2\theta(h_v + d) \qquad \text{(Eq. 7-8)}$$

7-14. Pressures underneath the gate will approach upstream pressure just prior to closure. The gate shown in Figure 7-22 is designed to allow this pressure to also act on top of the gate. This will assist in closing the gate.

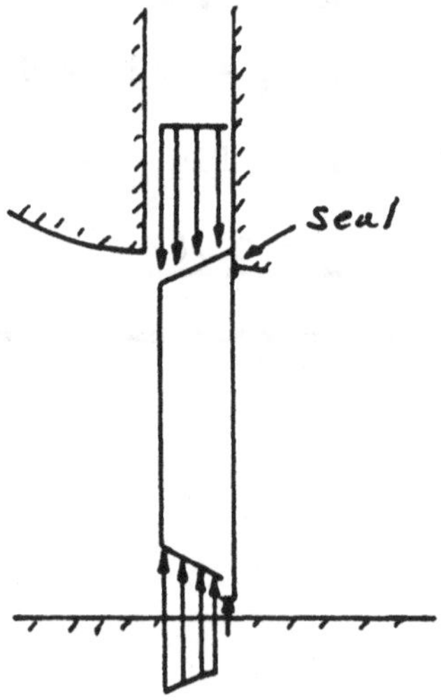

7-17. Flow during the pumping cycle will separate from the flow boundaries, with the result that velocities in the vicinity of the trashracks may not be substantially reduced from velocities in the penstock.

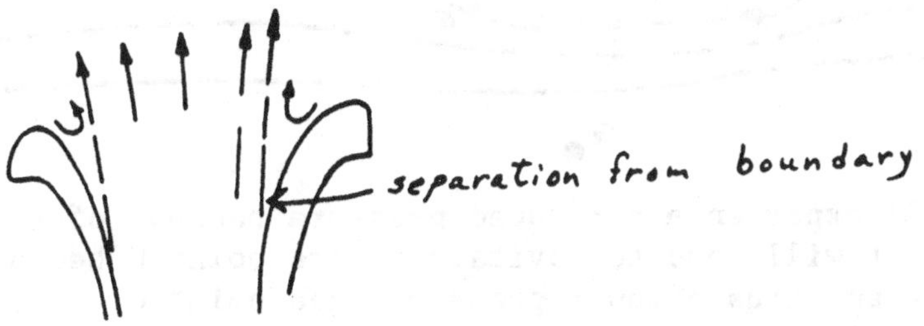

7-19. Given: 3 pumps

Q = 60,000 gpm

Then Q per pump = 20,000 gpm

Significant dimensions of the pump pit are picked from Fig. 7-39 as follows:

C = 25"
B = 37"
S = 82"
H = 88"
y = 140"
A = 225"

7-20. (1) Any increase in the ambient pressure by either increasing the downstream pressure or lowering the energy dissipator.

(2) Injection of air into the zone of separation.

(3) Treatment of surface with a material having high resistance to cavitation.

(4) Remove the boundary further from cavitation zone by increasing the size of the expansion.

(5) Restrict operation to a low level of cavitation so there is little energy in the cavitation and the cavities collapse before reaching the boundary.

7-21.

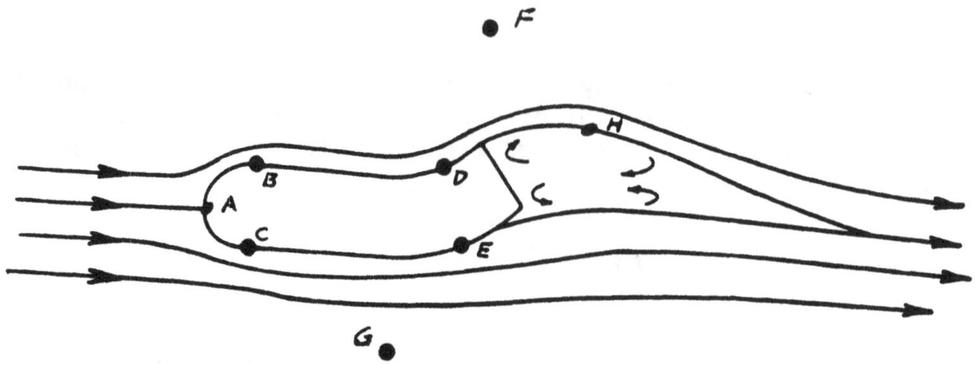

Point B will experience a reduced pressure because of the increase in velocity. It will tend to cavitate before point C because it has less submergence and thus a lower pressure than point C. There also may be early cavitation at point H, because it is in the zone of separation where pressures may be locally reduced in the cores of local vortices.

7-22. (1) At point A the velocity may not yet be high enough to cause cavitation.

(2) At point B, there may be cavitation because of the increased velocity, but the additional pressure generated by the concave upward curve will tend to delay cavitation until nearer the end of the curve.

(3) At point C, the velocities will be reduced near the boundary because of energy loss and re-adjustment of the velocity profile from the higher velocities established near the boundary by the vertical curve.

7-23. From Equation 7-14,

$$\sigma_i = 1 - 2\,C_p$$

$$= 1 - 2\,(1 - V_s^2/V_o^2)$$

$$= 1 - 2\,(1 - 9^2)$$

$$= 161$$

By Equation 7-9,

$$\sigma_i = (p_o - p_v)\,/\rho V_o^2/2$$

$$161 = (25(144) - 37)\ /\ (1.94\ V_o^2/2)$$

$$V_o = 4.78\ \text{fps}$$

The discharge

$$Q = V_o \times A_o$$

$$Q = 4.78 \times \pi/4 \times 2^2$$

Q = <u>15 cfs</u>

7-24. Velocities near the boundary are higher for a smooth boundary than a rough boundary. When these higher velocities are imposed on a rough surface downstream from a smooth surface, cavitation can occur because of the locally high velocities that occur as the water flows past the roughness elements.

7-25. from Bernoulli's equation,

$$\frac{p_o}{\gamma} + \frac{V_o^2}{2g} = \frac{p_s}{\gamma} + \frac{V_s^2}{2g}$$

$$p_s - p_o = \rho V_o^2/2\ (1 - V_s^2/V_o^2)$$

noting that $C_p = (p_s - p_o)/(\rho V_o^2/2)$,

$$C_p = 1 - V_s^2/V_o^2, \text{ and}$$

$$V_s = V_o\,(1 - C_p)^{1/2} \qquad \text{(Eq. 7-12)}$$

7-25. (continued).

From Equation 7-13, the pressure at the eddy center,

$p_{min} = p_s - \rho V_s^2/2$, and

$(p_{min} - p_o) / (\rho V_o^2/2) = (p_s - \rho V_s^2/2 - p_o) / (\rho V_o^2/2)$

Noting that $(p_{min} - p_o) / (\rho V_o^2/2) = -\sigma_i$ (separated flow)

and $p_s - p_o = C_p \rho V_o^2/2$,

σ_i (separated flow) $= (-C_p \rho V_o^2/2 + V_s^2/2) / (\rho V_o^2/2)$

σ_i (separated flow) $= -C_p + V_s^2/V_o^2$

σ_i (separated flow) $= 1 - 2\,C_p$ (Eq. 7-14)

7-26. From Equation 7-11, incipient cavitation

$\sigma_i = -C_p = (p_o - p_v) / (\rho V_o^2/2)$

$= (5195 - 37) / (1.94\,(100^2) / 2)$

$= 0.532$

Therefore, $C_p = -0.532$ when cavitation begins.

Thus, all of the upper side where C_p is between values of -0.532 and −1.2 will be cavitating.

7-27. Spillway flow is simulated by having the same Froude numbers in the model and prototype. The discharge ratio for this condition is given by Equation 7-26:

$Q_r = L_r^{5/2}$

Therefore, the prototype discharge is

$$Q_p = \frac{1}{Q_r} \times Q_m = (25)^{5/2} \times 0.1 = 312.5 \text{ m}^3/\text{s}$$

The time ratio $t_r = L_r / V_r = L_r^{1/2}$

7-27. (continued).

The prototype time required for a particle to traverse the corresponding distance a particle took 1 minute in the model to traverse is:

$$t_p = \frac{1}{t_r} \times 1 \text{ min}$$

$$= (25)^{1/2} \times 1 \text{ min}$$

$$= \underline{\underline{5 \text{ min}}}$$

7-28. Assuming model discharge is the governing factor, a maximum discharge ratio $Q_r = 0.9/5000$

$$= 1:5556$$

This corresponds to $L_r = Q_r^{0.4} = 31.5$

Choosing $L_r = 32$, model width and length requirements are $300/32 = 9.4$ m and $1200/32 = 37.5$ m. These can both be contained within the available space, so use

$$L_r = \underline{\underline{32}}$$

7-29. For incipient cavitation,

$$\sigma_i = -C_p = (p_o - p_v) / (\rho V_o^2/2)$$

Therefore, for $p_o = 5$ psig

$$-(-0.30) = (5(144) - 0.256\ (144)) / (1.94\ V_o^2/2)$$

$$V_o = 47.7 \text{ fps}$$

Thus, cavitation will occur if V_o = 50 fps.

If P_o = 10 psig,

$0.30 = (1440 - 37) / (1.94\ V_o^2/2)$

V_o = 67.3 fps for cavitation

Thus, if V_o = 70 fps, cavitation will occur.

7-30. (a) The mass ratio $M_r = \rho_r L_r^3 = (1)\ (1:16)^3$

$= 1:4096$

The model mass = 40,000 lbm / 4096

= 9.77 lbm

(b) The spring constant ratio $K_r = F_r / L_r$

$= L_r^3 / L_r$

$= L_r^2$

$= 1:256$

Therefore the model spring constant

K_m = 155,000 lb/in / 256 = 605 lb/in

(c) The frequency ratio $f_r = 1/t_r = 1/t_r = 1/L_r^{1/2}$

$= 4:1$

The prototype frequency f_p = 10 hz x 1/4 = 2.5 Hz

7-31. (a) Equation 7-29 and the choice of $\Delta\rho/\rho = 1$

results in the ordinary Froude number being satisfied, as well as the densimetric Froude number.

Thus, the ratio of heat transfer coefficients

$$K_r = y_r/t_r = y_r \times V_r \;/\; L_r = y_r^{3/2} \;/\; L_r$$

$$y_r = (L_r K_r)^{2/3} = (1{:}100 \times 1{:}3)^{2/3} = \underline{1{:}44.8}$$

(b) In a distorted model satisfying the Froude criterion, the discharge ratio

$$\begin{aligned} Q_r &= V_r A_r \\ &= y_r^{1/2} \times y_r \times L_r \\ &= y_r^{3/2} \times L_r \\ &= (1{:}44.8)^{3/2} \, (1{:}100) \\ &= 1{:}30{,}000 \end{aligned}$$

Therefore, the model flow

$$Q_m = 100{,}000 \;/\; 30{,}000 = \underline{3.33 \text{ cfs}}$$

(c) The Manning's ratio n_r is determined from Manning's equation, as follows:

$$V_r = \frac{1}{n_r} y_r^{2/3} \, S_r^{1/2}$$

$$y_r^{1/2} = \frac{1}{n_r} y_r^{2/3} \, y_r^{1/2} \;/\; L_r^{1/2}$$

$$n_r = y_r^{2/3} \;/\; L_r^{1/2} = \frac{(1{:}44.8)^{2/3}}{(1{:}100)^{\;1/2}} = \underline{1{:}1.26}$$

7-32. For enclosed systems of this type, it is desirable to operate the model and prototype at equal Reynolds number. Thus,

$$\left(\frac{Vd}{\nu}\right)_m = \left(\frac{Vd}{\nu}\right)_p, \text{ or}$$

$$V_m = V_p \, (d_p/d_m) \, (\nu_m/\nu_p)$$

$$= 10 \text{ m/s } (6) \, (3.26 \times 10^{-7} \;/\; 10.0 \times 10^{-7}) = 19.56 \text{ m/s}$$

7-32. (continued).

The corresponding model flow rate is

$$Q_m = V_m A_m = 19.56\ (\pi/4)\ (0.1/6)^2\ (1000)\ \text{liters/s}$$

$$= \underline{4.27\ \text{liters/s}}$$

With models of this type, velocities in the model usually exceed those in the prototype, and require large pumping capacity and power requirements. For this particular application, where the model throat velocity is nearly 20 m/s, cavitation is likely to occur in the throat section, unless the entire model is operated under a sufficiently high pressure.

Often it is not necessary to operate the model at prototype Reynolds number, although in this application, because of the streamlined shape of the Venturi meter, the prototype Reynolds number of 10^6 should be simulated in the model. If the shape being tested has sharp corners where flow separation is well defined, the model Reynolds number may not need to be as high in the model. For example, if a relatively small orifice in a pipe were being tested under the same conditions as for this Venturi meter, the model need only be operated at a Reynolds number of about 10^5. This can be done because the flow coefficient does not vary with Reynolds number above a value of 10^5.

7-33. $\rho_{w.w.} = 1.94\ \text{slugs/ft}^3$; $\rho_{s.w.} = 1.99\ \text{slugs/ft}^3$

$$B = g(\Delta\rho/\rho)Q = 32.2\ \text{ft/sec}^2\ (0.05/1.94)\ (30\ \text{ft}^3/\text{sec})$$

$$= 24.90\ \text{ft}^4/\text{sec}^2$$

Velocities:

$$w_m = 4.7\ (B/z)^{0.333} \qquad (7\text{-}37)$$

Then at z = 100 ft; x = 0 $\quad w_m = 4.7\ (24.90/100)^{0.333} = \underline{2.96\ \text{ft/sec}}$

at z = 200 ft; x = 0 $\quad w_m = 4.7\ (24.90/200)^{0.333} = \underline{2.35\ \text{ft/sec}}$

$$w = w_m\ e^{-100(x/z)} \qquad (7\text{-}38)$$

Then at z = 100'; x = 20' $\quad w = 2.96\ e^{-100(20/100)^2} = \underline{0.054}$ ft/sec.

at z = 200'; x = 20 $\quad w = 2.96\ e^{-100(20/200)^2} = \underline{1.089}$ ft/sec.

Dilutions: $S = 0.15\ B^{1/3}\ z^{5/3}/Q$

at z = 100 ft $\quad \bar{S} = 0.15\ (24.9)^{1/3}\ (100)^{5/3}/30 = \underline{31.45}$

at z = 200 ft $\quad S = 31.45\ (2)^{5/3} = \underline{99.8}$

7-34. $w = w_m\, e^{-87(x/z)}$ (7-31)

where $w_m = 6.2\, w_o(D/z)$ (7-30)

or $w = 6.2\, w_o\, (D/z)\, e^{-87(x/z)}$

For this problem $w_o = Q/A = 30 / ((\pi/4)1^2) = 38.2$ ft/sec

$D = 1$ ft

Results are shown in table (below)

x(ft)	z(ft)	D/z	x/z	w(ft/sec)
0	20	1/20	0	11.84
0	40	1/40	0	5.92
6	20	1/20	6/20	0.005 ≈ 0
6	40	1/40	6/40	0.84

==

7-35. $z_{max} = 3.6\, (g'q)^{1/3}\, [(-g/\rho)\, (d\rho_a/dz)]^{-1/2}$ (7-44)

Note: assume $\rho_{wastewater} = 997$ kg/m^2 for T = 75° F = 23.9° C

where $g' = g(\Delta\rho/\rho) = 9.81$ m/s^2 $((1024 - 997)/1024) = 0.259$ m/s^2

$d\rho_a/dz \approx (1023 - 1024)/70 = -0.0143$ kg/m^3/m

$q = 7$m^3/s / 1030 m = 0.0068 m^2/s

Then $z_{max} = 3.6\, (0.259 \times 0.0068)^{1/3}\, [(-9.81/1024)\, (-0.0143)]^{-1/2}$

$= \underline{37 \text{ m}}$

$S_{min} = 0.24\, (g')^{1/3}\, z_{max}/q^{2/3}$ (7-45)

$= 0.24\, (0.259)^{1/3}\, (37)\, /\, (0.0068)^{2/3}$

$= \underline{157}$

==

7-36. Determine the downstream depth y_3 from Manning's equation.

$Q = by\ (1.49/n)(y_3^{2/3})(S)^{1/2}$

$150{,}000 = 600(1.49/0.02)(y_3^{5/3})(0.001)^{1/2}$

Solving, $y_3 = 16.4$ ft

$V_3 = Q/A_3 = 150{,}000/16.4(600) = 15.2$ ft/s

The downstream Froude No.

$Fr_3 = V_3/(gy_3)^{1/2}$

$= 15.2/(32.2 \times 16.4)^{1/2}$

$= 0.66$

Compute the drop in water surface Δy through the bridge piers by using Yarnell's equation:

$\Delta y/y_3 = KFr_3^2(K + 5Fr_3^2 - 0.6)(\alpha + 15\alpha^4)$

where

$K = 0.90$ from Table 7-1

$= (b_1 - b_2)/b_1 = (60-50)/60 = 0.167$

Therefore

$\Delta y/y_3 = 0.90(0.66)^2(0.9 + 5(0.66)^2 - 0.6)(0.167 + 15(0.167)^4)$

$= 0.174$

Therefore $\Delta y = 0.174(16.4) =$ <u>2.85 ft</u>

7-37. The unit discharge

$Q = 200{,}000/150 = 1{,}333$ cfs/ft

$= 123.8\ m^3/s/m$

$d = 1$ ft $= 0.305$ m

$H = 2{,}000 - 1{,}820 = 180$ ft

$= 54.86$ m

$h = 1{,}820 - 1{,}790 = 30$ ft $= 9.14$ m

The depth of scour, measured in meters from the tailwater level, given by the Veronese equation is

$D = 1.9H^{0.225}q^{0.54}$

$= 1.9(54.86)^{0.225}(123.8)^{0.54}$

$= 63.1$ m

$= \underline{207\ \text{ft}}$

The depth of scour in meters can also be computed from the equation developed by Mason and Arumugam:

$D = Kq^xH^yh^w/(g^vd^z)$, where

$K = 6.42 - 3.10\ H^{0.10} = 6.42 - 3.10 \times 54.86^{0.10}$

$= 1.79$

$v = 0.30$

$w = 0.15$

$x = 0.60 - H/300 = 0.60 - 54.86/300 = 0.417$

$y = 0.05 + H/200 = 0.05 + 54.86/200 = 0.324$

$z = 0.10$

$D = 1.79(123.8)^{0.417}(54.86)^{0.324}(9.14)^{0.15}/((9.81)^{0.30} \times (0.305)^{0.10})$

$\underline{= 38.7\ \text{m}}$

$\underline{= 127\ \text{ft}}$

The substantial difference in depth predicted by the two equations indicates the lack of accuracy in predicting plunge pool scour. Equations of this type should be used only for preliminary design. An extensive understanding of the plunge pool geology, and perhaps the use of a model study, are needed for more accurate prediction.

CHAPTER EIGHT

8-1. $D = 40$ cm; $n = 1{,}000/60 = 16.67$ rev/sec; $\Delta h = 3$ m

$C_H = \Delta hg/D^2n^2 = 3 \times 9.81/((0.4)^2 \times (16.67)^2) = 0.662$

from Fig. 8-5, $C_Q = Q/(nD^3) = 0.625$

Then $Q = 0.625 \times 16.67 \times (0.4)^3 = \underline{0.667\ m^3/s}$

8-2. $n = 690/60 = 11.5$ rev/sec; $D = 71.2$ cm; $\Delta h = 10$ m

$C_H = \Delta hg/(n^2D^2) = 10 \times 9.81/((0.712)^2(11.5)^2) = 1.46$

from Fig. 8-5, $C_Q = 0.40$ and $C_P = 0.76$

$Q = C_Q nD^3 = 0.40 \times 11.5 \times 0.712^3 = \underline{1.66\ m^3/s}$

Power $= C_P \rho D^5 n^3 = 0.76 \times 1{,}000 \times 0.712^5 \times 11.5^3 = \underline{211\ kW}$

8-3. At maximum efficiency, from Fig. 8-5,

$C_Q = 0.64$; $C_P = 0.60$; and $C_H = 0.75$

$D = 2$ ft; $n = 1{,}750/60 = 29.17$ rev/sec

$Q = C_Q nD^3 = 0.64 \times 29.17 \times 2^3 = \underline{149.4\ cfs}$

$\Delta h = C_H n^2 D^2/g = 0.75 \times 29.17^2 \times 2^2/32.2 = \underline{79.3\ ft}$

Power $= C_P \rho D^5 n^3 = 0.60 \times 1.94 \times 2^5 \times 29.17^3 = 924{,}510$ ft-lb/sec $= \underline{1{,}681\ hp}$

8-4. At maximum efficiency, from Fig. 8-5,

$C_Q = 0.64$; $C_P = 0.60$: $C_H = 0.75$; $D = 0.50$ m; $n = 30$ rps

$Q = C_Q nD^3 = 0.64 \times 30 \times 0.5^3 = \underline{2.40\ m^3/s}$

$\Delta h = C_H n^2 D^2/g = 0.75 \times 30^2 \times 0.5^2/9.81 = \underline{17.2\ m}$

Power $= C_P \rho D^5 n^3 = 0.60 \times 1{,}000 \times 0.5^5 \times 30^3 = \underline{506\ kW}$

8-5. $D = 14/12 = 1.167$ ft; $n = 900/60 = 15$ r/s

$\Delta h = C_H n^2 D^2/g = C_H (15)^2 (1.167)^2/32.2 = 9.52\ C_H$ ft

$Q = C_Q n D^3 = C_Q 15(1.167)^3 = 23.8\ C_Q$ cfs

C_Q	C_H	Q(cfs)	Δh(ft)
0.0	2.9	0.0	27.60
0.1	2.55	2.38	24.27
0.2	2.0	4.76	19.04
0.3	1.7	7.15	16.18
0.4	1.5	9.53	14.28
0.5	1.2	11.91	11.42
0.6	0.85	14.29	8.09

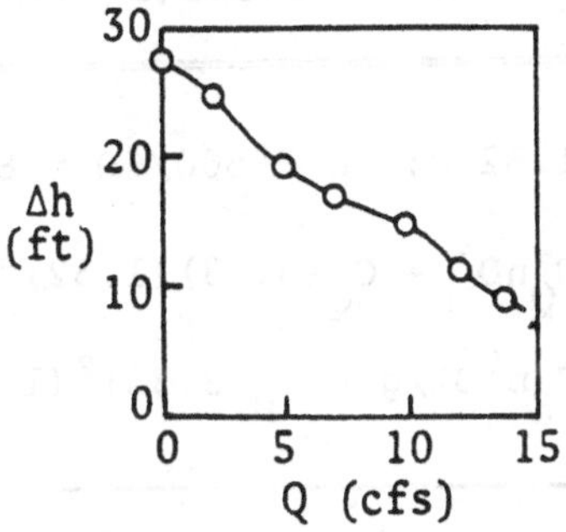

8-6. $D = 60$ cm $= 0.60$ m; $N = 690$ rpm $= 11.5$ rev/sec

Then $\Delta h = C_H D^2 n^2/g = 4.853\ C_H$

$Q = C_Q n D^3 = 2.484\ C_Q$

C_Q	C_H	$Q(m^3/s)$	h(m)
0.0	2.90	0.0	14.1
0.1	2.55	0.248	12.4
0.2	2.00	0.497	9.7
0.3	1.70	0.745	8.3
0.4	1.50	0.994	7.3
0.5	1.20	1.242	5.8
0.6	0.85	1.490	4.2

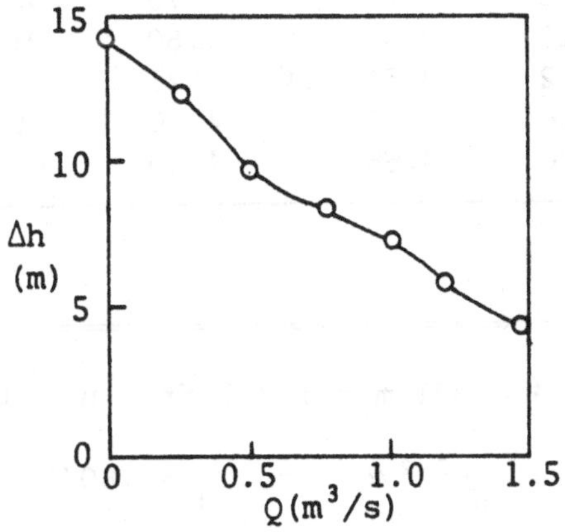

8-7. $H \alpha n^2$ so $H_{30}/H_{35.6} = (30/35.6)^2$

or $H_{30} = 104 \times (30/35.6)^2 = \underline{73.8\ m}$

8-8. $D = 0.40$ m $\quad n = 30$ rps

$C_H = \Delta h g/(n^2 D^2) = 50(9.81)/[(30)^2(0.40)^2] = 4.91$

from Fig. 8-9 $\quad C_Q = 0.136$

then $Q = C_Q n D^3 = 0.136(30)(0.40)^3 = \underline{0.261\ m^3/s}$

8-9. $D = 0.371 \times 2 = 0.742$ m; $n = 2{,}133.5/(2 \times 60) = 17.77$ rps

from Fig. 8-8, at peak efficiency $C_Q = 0.121$, $C_H = 5.15$

$\Delta h = C_H n^2 D^2/g = 5.15(17.77)^2(0.742)^2/9.81 = \underline{\underline{91.3 \text{ m}}}$

$Q = C_Q nD^3 = 0.121(17.77)(0.742)^3 = \underline{\underline{0.878 \text{ m}^3/\text{s}}}$

8-10. $D = 1.52$ m; $n = 500/60 = 8.33$ rps

$Q = C_Q nD^3 = C_Q(8.33)(1.52)^3 = 29.27\ C_Q \text{ m}^3/\text{s}$

$\Delta h = C_H n^2D^2/g = C_H(8.33)^2(1.52)^2/9.81 = 16.36\ C_H \text{ m}$

C_Q	Q	C_H	Δh
0.0	0.0	5.80	94.9
0.04	1.17	5.80	94.9
0.08	2.34	5.75	94.1
0.10	2.93	5.60	91.6
0.12	3.51	5.25	85.9
0.14	4.10	4.80	78.5
0.16	4.68	4.00	65.4

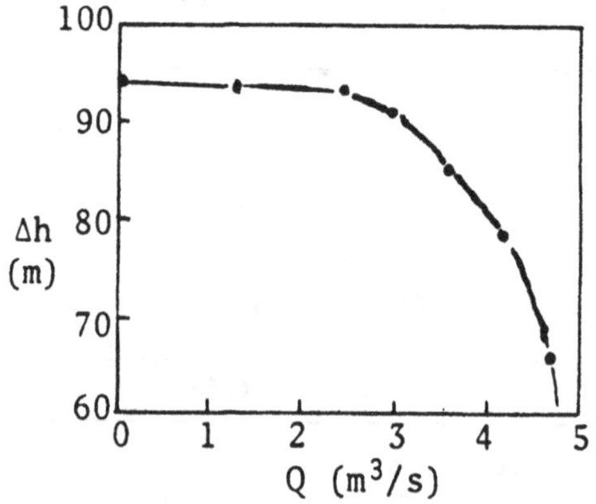

8-11. $D = 0.371 \text{ m} = 1.217$ ft; $n = 1{,}750/60 = 29.17$ rps

$\Delta h = C_H n^2D^2/g$ so $C_H = 160(32.2)/[(29.17)^2(1.217)^2] = 4.09$

from Fig. 8-8 $C_Q = 0.105$

Then $Q = C_Q nD^3 = 0.105(29.17)(1.217)^3 = \underline{\underline{5.52 \text{ cfs}}}$

8-12. $C_H = \Delta Hg/D^2n^2$ Since C_H will be the same for the maximum head condition, then $\Delta H \propto n^2$ or $H_{1,500} = H_{1,000} \times (1{,}500/1{,}000)^2$

$H_{1,500} = 102 \times 2.25 = \underline{\underline{229.5 \text{ ft}}}$

8-13. Necessary conversion factors:

1 BTU = 778 ft-lb

1 horsepower = 550 ft-lb/s

Also Specific heat of water = 1 BTU / lbm • °F

The mass rate of flow through pump is given as:

$$\dot{m} = \rho Q$$

where $\dot{m}$ = mass rate in lbm/s

ρ = density in lbm/ft^3

Q = discharge in ft^3/s

If we let the water temperature rise 15°F as it flows through the pump, then the BTU's added each second will be:

$$\text{BTU rate} = 15\rho Q \qquad (1)$$

or the energy rate added in ft·lb/s will be

$$\text{Energy rate} = \text{BTU rate x 778 ft·lb/BTU} \qquad (2)$$

$$HP = \text{Energy rate}/550 \qquad (3)$$

where HP = horsepower

Solving Eqs. (1), (2) and (3) for Q yields

$$Q = 550\, P/(15 \times 778\rho) \qquad (4)$$

If ρ = 62.4 lbm/ft

Then Q = 550 P/(15 x 778 x 62.4)

<u>**$Q = 7.55 \times 10^{-4} HP$**</u>

Where Q is in ft^3/s and HP is in horsepower.

Note: In the above derivation it was assumed that the power added to the flow is essentially the same as the power for shutoff conditions. In most cases this is valid assumption because the P-Q curve is essentially flat at the very low flow condition.

8-14. Shut off power for the pump of Fig. 8-8 is approximately 93 kW.

Power in horsepower $= 93/0.747$

$= 124.5$

To limit the temperature rise to $15^\circ F$ use Eq. (8-11):

$$Q = 7.55 \times 10^{-4} P$$

$$= 7.55 \times 10^{-4} \times 124.5$$

$$= \underline{\mathbf{0.094\ cfs}}$$

8-15. Assume that all the power given in Fig. 8-8 is used to either pump the water or heat the water; none is used in bearing friction. The power for pumping is given by Eq. (5-3):

$$P = Q\gamma h_p$$

where $Q = 0.094$ cfs (from solution to Prob. 8-14)

$\gamma = 62.4\ lb/ft^3$

$h_p = 108$ m (scaled from Fig. 8-8)

$= 354$ ft

Then $P = 0.094 \times 62.4 \times 354$

$= 2{,}076$ ft-lb/s

$= 3.77$ horsepower

The shutoff power is 93 kW = 124.5 horsepower

Therefore, (3.77/124.5) x 100 = **3 %** of the power is actually used for pumping and **97%** is used in heating the water.

8-16. $\Delta T = 212^{\circ} - 60^{\circ} = 152\,^{\circ}F$

Mass of water = 1 ft^3 x 62.4 lbm/ft^3

= 62.4 lbm

Heat that must be added to produce boiling

= 152 °F x 62.4 lbm x specific heat

= 152 °F x 62.4 lbm x 1 BTU/lbm • °F

= 9,485 BTU

But 1 BTU = 778 ft-lb

Therefore; ft-lb to be added for boiling

= 778 x 9,485

= 7,379 x 10^6 ft-lb

From solution to Prob. 8-14 the horsepower at shutoff = 124.5 horsepower or the ft-lb of energy added to the flow is 124.5 x 550 = 68,475 ft-lb/s

Thus 68,475 ft lb/s x t = 7.379 x 10^6 ft-lb

t = **108 sec**

8-17. c_p = Specific heat of water = 1 cal./(g· °C)

= 4,187 J/(kg· °C)

= 4,187 J/(kg· °C)

= 4.187 kJ/(kg· °C)

If water is flowing at ρQ kg/s, then the power required to heat the water ΔT° will be

$P = \rho Q \Delta T \times 4.187$

where P = power in kW

or $Q = P/(4.187 \rho \Delta T)$

8-17. (Continued)

For this problem

$$Q = 93 \text{ kW} / ((4.187 \text{ kJ/kg}\cdot{}^\circ\text{C})(1{,}000 \text{ kg/m}^3) \times (10\ {}^\circ\text{C}))$$

$$= (93 \text{ kJ/s}) / ((4.187 \text{ kJ} / (\text{kg}\cdot{}^\circ\text{C}))(1{,}000 \text{ kg/m}^3)(10\ {}^\circ\text{C}))$$

$Q = 0.00222\ \text{m}^3/\text{s}$

8-18. $c_p = 4.187 \text{ kJ} / (\text{kg}\cdot{}^\circ\text{C})$

Volume of water being heated = $0.1\ \text{m}^3$

Then mass of water being heated $= \rho \forall = 0.1\rho$

Power to pump = 93 kW

= 93 kJ/s

Energy that must be supplied to $0.1\ \text{m}^3$ of water to heat it $10\ {}^\circ\text{C}$ will be

$$E = M c_p$$

$$= \rho \forall c_p \Delta T$$

$$Pt = \rho \forall c_p \Delta T$$

where t = time to heat water

Then $t = \rho \forall c_p / (P)$

$$= (10{,}00 \text{ kg/m}^3)(0.1\ \text{m}^3)(4.187 \text{ kJ} / (\text{kg}\cdot{}^\circ\text{C})(10\ {}^\circ\text{C}) / 93 \text{ kW}$$

$$= (1000 \text{ kg/m}^3)(0.1\ \text{m}^3)(4.187 \text{ kJ/kg}\cdot{}^\circ\text{C})(10\ {}^\circ\text{C}) / (93 \text{ kJ/s})$$

= 45 sec

8-19. N = 1,500 rpm so n = 25 rps; Q = 12 cfs; h = 25 ft

$$n_s = n\sqrt{Q}/[g^{3/4}h^{3/4}] = (25)(12)^{1/2}/[(32.2)^{3/4}(25)^{3/4}] = 0.57$$

Then from Fig. 8-15, $n_s < 0.60$, so use a mixed flow pump.

8-20. $n = 25$ rps; $Q = 0.30\ m^3/sec$; $h = 8$ meters

$n_s = n\sqrt{Q}/[g^{3/4}h^{3/4}] = (25(0.3)^{1/2}/[(9.81)^{3/4}(8)^{3/4}] = 0.52$

Then from Fig. 8-15, $n_s < 0.60$ so <u>use a mixed flow pump</u>.

8-21. $N = 1{,}100$ rpm $= 18.33$ rps; $Q = 0.4\ m^3/sec$; $h = 70$ meters

$n_s = n\sqrt{Q}/[g^{3/4}h^{3/4}] = (18.33)(0.4)^{1/2}/[(9.81)^{3/4}(70)^{3/4}]$

$= (18.33)(0.63)/[(5.54)(24.2)] = 0.086$

Then from Fig. 8-15, $n_s < 0.23$ so <u>use a radial flow pump</u>.

8-22. $N = 1{,}100$ rpm $= 18.33$ rps; $Q = 12$ cfs; $h = 600$ ft

$n_s = n\sqrt{Q}/[g^{3/4}h^{3/4}] = (18.33)(12)^{1/2}/[(32.2)^{3/4}(600)^{3/4}]$

$= (18.33)(3.46)/[(13.5)(121)] = 0.039$

Then from Fig. 8-15, $n_s < 0.23$, so <u>use a radial flow pump</u>.

8-23. $n_s = n\sqrt{Q}/(g^{3/4}h^{3/4})$; $n = 10$ rps; $Q = 1.0\ m^3/s$; $h = 3 + (1.5+fL/D)V^2/(2g)$;

$V = 1.27$ m/s Assume $f = 0.01$, so

$h = 3 + (1.5 + 0.01 \times 20/1)(1.27)^2/(2 \times 9.81) = 3.14$ m

Then $n_s = 10 \times \sqrt{1}/(9.81 \times 3.14)^{3/4} = 0.76$

From Fig. 8-15, <u>use axial flow pump.</u>

8-24. $D = 35.6$ cm; $n = 11.5$ r/s

Writing the energy equation from the reservoir surface to the center of the pipe at the outlet,

$p_1/\gamma + V_1^2/(2g) + z_1 + h_p = p_2/\gamma + V_2^2/(2g) + z_2 + \Sigma h_L$

$h_p = 21.5 + 20 + [Q^2/(A^2 2g)](1 + fL/D + k_e + k_b)$

$L = 64$ m assume $f = 0.014$ $r_b/D = 1$

from Table 5-3, $k_b = 0.35$ $k_e = 0.1$

(1) $h_p = 1.5 + [Q^2((0.014(64)/0.356) + 0.35 + 0.1 + 1)]/[2(9.81)(\pi/4)^2(0.356)^4] = 1.5 + 20.42Q^2$

$C_Q = Q/(nD^3) = Q/[(11.5)(0.356)^3] = 1.93Q$

(2) $h_p = C_H n^2 D^2/g = C_H(11.5)^2(0.356)^2/9.81 = 1.71C_H$

8-24. (Continued)

Q(m/s)	C_Q	C_H	$h_p(1)$ (m)	$h_p(2)$ (m)
0.10	0.193	2.05	1.70	3.50
0.15	0.289	1.70	1.96	2.91
0.20	0.385	1.55	2.32	2.65
0.25	0.482	1.25	2.78	2.13
0.30	0.578	0.95	3.34	1.62
0.35	0.675	0.55	4.00	0.94

Then plotting the system curve and the pump curve, we obtain the operating condition:

$Q = 0.21\ m^3/s$;   Power = 6.7 kW (from Fig. 8-6)

Graph for solution of Problems 8-24 & 8-25:

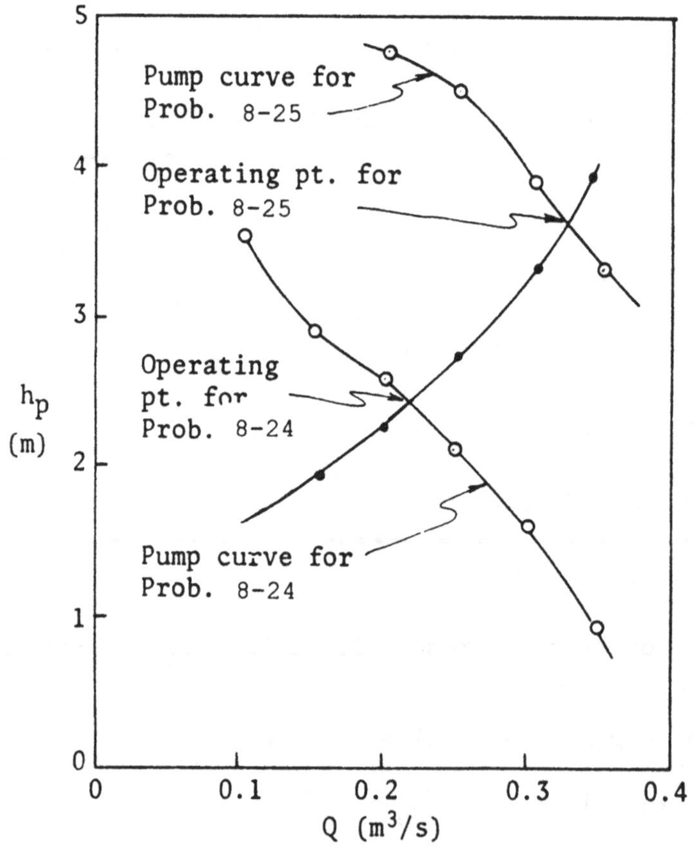

8-25. The system curve will be the same as in Problem 8-24.

$C_Q = Q/[nD^3] = Q/[15(0.356)^3] = 1.48Q$

$h_p = C_H n^2 D^2/g = C_H (15)^2 (0.356)^2/9.81 = 2.91C_H$

Q	C_Q	C_H	h_p
0.20	0.296	1.65	4.79
0.25	0.370	1.55	4.51
0.30	0.444	1.35	3.92
0.35	0.518	1.15	3.34

Plotting the pump curve with the system curve gives the operating condition:

$Q = \underline{0.32\ m^3/s}$; $C_Q = 1.48(0.32) = 0.474$

Then from Fig. 8-5, $C_p = 0.70$

Power $= C_p n^3 D^5 \rho = 0.70(15)^3(0.356)^5 1{,}000 = \underline{13.5\ kW}$

8-26. Assume atmospheric pressure = 101 kN/m^2

Assume temperature = 10° C; $p_V = 1.2\ kN/m^2$

Then NPSH = 1.5 m + $(101\ kN/m^2 - 1.2\ kN/m^2)/9.81\ kN/m^3$

= 11.7 m

Assume the critical specific speed is applicable to axial flow pumps as well as to the conventional centrifugal pumps.

Therefore,

$n_{ss,crit.} = 0.494 = n\sqrt{Q}/(g^{3/4} \times NPSH^{3/4})$

or $n = 0.494 \times g^{3/4} \times NPSH^{3/4}/\sqrt{Q}$

$n = 0.494 \times 9.81^{3/4} \times 11.7^{3/4}/\sqrt{0.4}$

= 24.4 rev/s

= $\underline{1{,}466\ rpm}$

8-27. $N = 690$ rpm = 11.5 rps; $Q = 0.21\ m^3/s$

Assume temperature = 10° C; $p_v = 1.2\ kN/m^2$

Vapor pressure head = $p_v/\gamma = 1.2/9.81 = 0.12$m

Assume atmospheric pressure head 10.3m

Neglecting head loss and velocity head, the gage pressure head on the suction side of the impellor will be approximately 1 m (see Prob. 8-24).

Then NPSH = 10.3m + 1 m - 0.12 = 11.2m

$$n_{ss} = nQ^{1/2}/(g^{3/4} \times NPSH^{3/4})$$
$$= 11.5 \times 0.21^{1/2}/(9.81^{3/4} \times 11.2^{3/4})$$
$$= 11.5 \times 0.458/(5.543 \times 6.12)$$
$$= 0.155$$

The n_{ss} value of 0.155 is much less than the critical value of 0.494; therefore, the pump is in the safe operating range.

Note: This assumes that the critical specific speed of 0.494 is the same as given by the Hydraulic Institute for conventional centrifugal pumps.

8-28. **$D = 0.36$ m; $L = 610$ m; $z = 450 - 366 = 84$ m**

Assume $\Delta h = 90$ m [$>\Delta z$], then from Fig. 8-8, $Q = 0.24\ m^3/s$

$V = Q/A = 0.24/[(\pi/4)(0.36)^2] = 2.36$ m/s; $k_s/D = 0.00012$

Assuming T = 20°C, $Re = VD/\nu = 2.36(0.36)/10^{-6} = 8.5 \times 10^5$

from Fig. 5-4, f = 0.014

$h_f = (0.014(610)/0.36)\ (2.36)^2/(2 \times 9.81)) = 6.73$ m; $h \approx 84+6.7 = 90.7$m

from Fig. 8-8, $Q = 0.23 m^3/s$; $V = 0.23/((\pi/4)(0.36)^2) = 2.26$ m/s

$h_f = [0.014(610)/0.36]\ (2.26)^2/(2 \times 9.81) = 6.18$ m

so $h = 84 + 6.2 = 90.2$ m and from Fig. 8-8 $Q = 0.225\ m^3/s$

8-29. The hydraulic grade line and energy grade line for the system will be shown below. Thus the minimum pressure point will be in the pipe just upstream of the pump. One can determine the magnitude of the pressure if one knows the discharge.

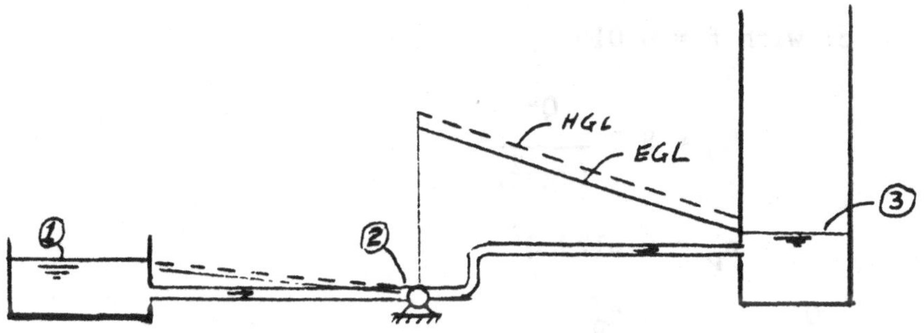

The discharge has to be determined by writing the energy equation from the reservoir to the tank and incorporating the pump characteristics:

$$p_1/\gamma + V_1^2/2g + z_1 + h_p = p_3/\gamma + V_3^2/2g + z_3 + \Sigma h_L \quad (0)$$

where $p_1/\gamma = 0$

$V_1^2/2g = 0$

$z_1 = 100$ ft

$p_3/\gamma = 0$

$V_3^2/2g = 0$

$z_2 = 100$ ft

$$\Sigma h_L = (K_{entrance} + 2K_{bends} + K_{outlet} + \frac{fL}{D})\, V^2/2g$$

Assume $K_{entrance} = 0.50$

$K_{bend} = 0.35$ from Table 5-3.

$K_{outlet} = 1.00$

Then $h_p = (0.5 + 2 \times 0.35 + 1.0 + f\ (800/2))\ (Q^2/A^2)/(2g)$

Assume $f = 0.015$; $A = (\pi/4) \times 2^2 = \pi\ ft^2$

8-29 (continued).

Then $h_p = (2.2 + 400f) \dfrac{Q^2}{(2g\pi^2)}$ (1)

or with f = 0.015

$$h_p = 8.2 \frac{Q^2}{(2g\pi^2)}$$

$$h_p = 0.0129\, Q^2 \qquad (2)$$

Q	h_p
30 cfs	11.6 ft
40	20.6 ft

Then solving Eq. (2) with the given performance curve yields an initial Q of 41 cfs (V = Q/A = 13.05 ft/s)

Now check f: $R = VD/\nu = \dfrac{(13.05 \times 2)}{(1.22 \times 10^{-5})} = 2 \times 10^6$

from Figs. 5-4, 5-5

f = 0.012

Then a more accurate equation for h_p would be

$$h_p = (2.2 + 400 \times 0.12) \frac{Q^2}{(2g\pi^2)}$$

$$h_p = 0.011\, Q^2 \qquad (3)$$

The discharge Q is still <u>about 41 cfs</u>

Q	hp
40 cfs	17.6 ft
41	18.5 ft

Now use this Q to get the minimum pressure.

8-29 (continued).

Write the energy equation from (1) to (2).

$$\frac{p_1}{\gamma} + \frac{V_1^2}{2g} + Z = \frac{p_2}{\gamma} + \frac{V_2^2}{2g} + Z_2 + (K_{entrance} + (\frac{fL}{D}))\frac{V^2}{2g}$$

$$100 = \frac{p_2}{\gamma} + 2.64 + 60 + (0.5 + \frac{0.012 \times 200}{2}) \times 2.64$$

$$\frac{p_2}{\gamma} = 100 - 2.64 - 4.49$$

$$= 32.9 \text{ ft}$$

$$p_2 = 32.9 \times 62.4 = 2051 \text{ psf gage}$$

$$= \underline{14.24 \text{ p.s.i. gage}}$$

b.

$$\text{Power} = \frac{Q\gamma h_p}{550 \times \text{Eff.}}$$

$$= \frac{41 \text{ ft}^3/\text{sec} \times 62.4 \text{ lb/ft}^3 \times 18.5 \text{ ft}}{550 \times 0.7}$$

Power = <u>123 horsepower</u>

c.

Assume f is constant at f = 0.012

Then $h_p = 0.011Q^2$

Determine the time to fill by taking the filling process in 20 foot steps and adjusting Q for each step.

z_3 $\quad$ $\bar{z}_3$ $\quad$ $\bar{Q}^*$ $\qquad$ $\Delta t = \frac{\forall}{Q} = \frac{\pi/4 \times 100^2 \times 20'}{Q}$

100'			
	110'	40 cfs	3927 sec
120'			
	130'	38 cfs	4133 sec
140'			
	150'	36 cfs	4363 sec
160'			
	170'	33 cfs	4760 sec
180'			
	190'	29 cfs	5416 sec
200'			

$t = 22{,}600$ sec

$\underline{t = 6.28 \text{ hours}}$

*To solve for Q we solve Eq. (0) from the reservoir to the tank with $z_3 = \bar{z}$ or

$$100' + h_p = \bar{z} + 0.11Q^2 \text{ (assuming } f = 0.012)$$

$$h_p = (\bar{z} - 100) + 0.011\,Q^2 \qquad (4)$$

8-30. The system Eq. is $h_p = 0.011\,Q^2$(Eq. (3) of Sol. to 8-29)

The Δh - Q curve for the pumps operating in parallel is given below:

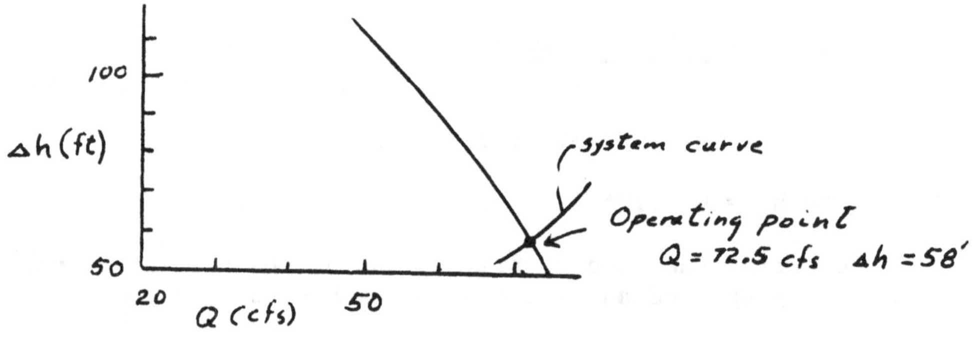

<u>Q = 72.5 cfs</u>

8-31. Assume the same system equation as for problem 8-29: $h_p = 0.011Q^2$.

The Δh - Q curve for the series operations is given below:

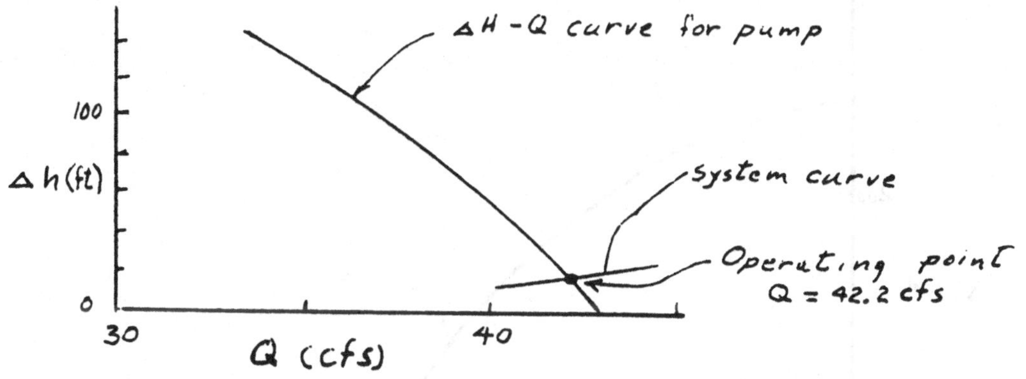

Q = 42.2 cfs

8-32. First write the energy equation from the lower to upper reservoir:

$$p_1/\gamma + V_1^2/2g + z_1 + h_p = p_2/\gamma + V_2^2/2g + z_2 + \Sigma h_L$$

$$0 \quad + \quad 0 \quad + 95 + \quad h_p = 0 \quad + \quad 0 \quad + z_2 + \Sigma h_L$$

$$h_p = (z_2 - 95) + (K_e + 2K_B + K_E + fL/D)\,V^2/2g$$

$K_e = 0.1$ (assumed); $K_b = 0.2$ (assumed for $r/d \approx 2$)

$K_E = 1.0$; $k_s/D = 0.0001$ (Fig. 5-5)

Assume $f = 0.014$ (Fig. 5-4)

Then $h_p = (z_2 - 95) + (0.1 + 2\,(.2) + 1.0 + .014\,(300/(1.5))\,V^2/2g$

$$h_p = (z_2 - 95) + 4.30\,V^2/2g$$

$$= (z_2 - 95) + 4.30\,Q^2/(2gA^2); \qquad A = (\pi/4)\,1.5^2 = 1.767\ \text{ft}^2$$

$$= (z_2 - 95) + 0.0214\,Q^2$$

The performance curve for the two pumps in series is given below.

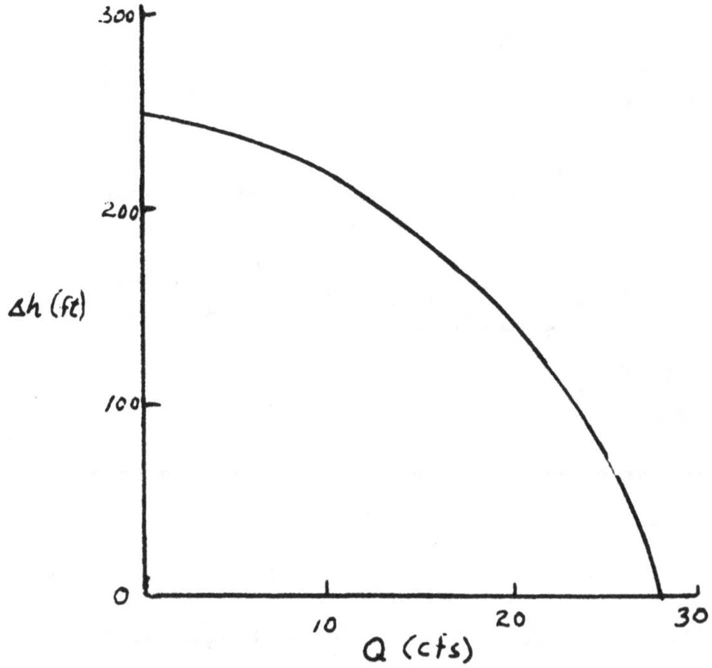

The initial discharge will be obtained by solving the performance curve and the energy Eq. (for $z_2 = 150$ ft):

$h_p = 55 + 0.0214\,Q^2$

$Q_i = 25.5$ cfs

To calculate the time to fill the reservoir consider increments of filling in 10 ft steps. The tabulation is shown below.

z_2(ft)	$\bar{z}_2$(ft)	$\bar{Q}^*$	$\Delta V(ft^3)$	Δt(sec)
150				
	155	25.4	50,265	1979
160				
	165	25.0	↓	2010
170				
	175	24.0		2094
180				
	185	23.5		2139
190				
	195	22.0	↓	2285
200				

$\Delta V = (\pi/4)\,80^2 \times 10 = 50{,}265\ ft^2$

$\Delta t = V/\bar{Q}$

$\Sigma\Delta t = t = 10{,}507$ sec $=$ **2.92 hours**

8-32 (continued).

*Note: Q is obtained by solving the system equation with the performance curve as done for obtaining Q_i.

Check f: $V = Q/A = 25.5/1.767 = 14.43$ ft/s

$R_e = VD/\nu = 14.43 \times 1.5/(1.2 \times 10^{-5}) = 1.8 \times 10^6$

$f = 0.013$

Based upon the Q_i the initial f value is slightly smaller than the assumed value. However, f will become larger as the tank fills (R_e will become smaller); therefore, use f = 0.014 for all calculations.

The maximum pressure will occur immediately downstream of the pumps when the 200 foot level is reached in the tank. Write the energy equation from the maximum pressure point to the water surface in the reservoir.

$$p_{max}/\gamma + V^2/2g + z = p_{res}/\gamma + V^2/2g + z_{res.} + \Sigma h_L$$

$$V = 22/(1.767) = 12.45 \text{ ft/sec}$$

$$p_{max}/\gamma + 2.41 + 90 = 0 + 0 + 200 + (1 + .014\ (280/1.5))V^2/2g$$

$$p_{max} = \gamma[(200-90) - 2.41 + 3.61\ V^2/2g]$$

$$= 62.4\ (110 - 2.41 + 3.61 \times 2.41)$$

$$p_{max} = 62.4\ (116.29) = \underline{7257 \text{ psfg}} = \underline{50.4 \text{ psig}}$$

Consider parallel pump installation:

The performance curve for the two pumps would be as shown below:

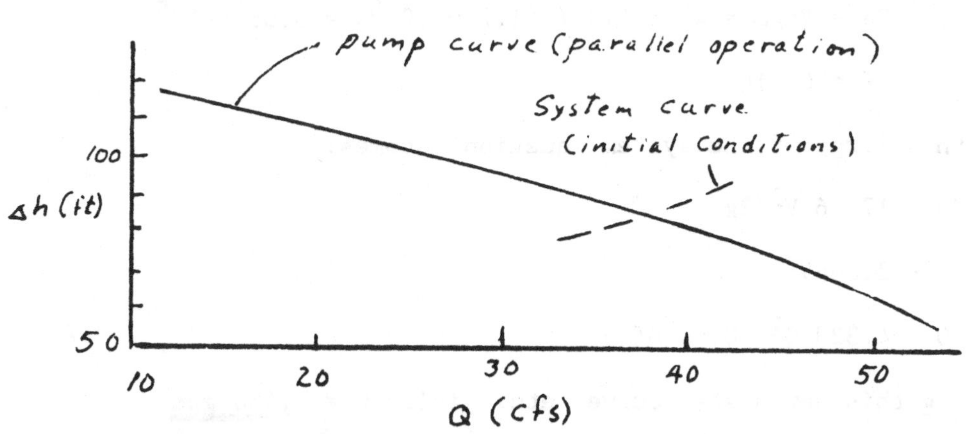

Solving the initial system equation with this performance curve yields Q_i = <u>39 cfs</u>

8-33. Write the energy equation from the water surface of reservoir A to the water surface of reservoir B:

$$p_1/\gamma + V_1^2/2g + z_1 + h_p = p_2/\gamma + V_2^2/2g + z_2 + \Sigma h_L$$

$$0 + 0 + 0 + h_p = 0 + 0 + 30 + \Sigma h_L$$

$$h_p = 30 + (K_e + K_E + 2K_b + 2K_V + fL/D)\, V^2/2g$$

where $K_e = 0.5$, $K_E = 1.0$, $K_b = 0.35$, $K_V = 0.20$ (Table 5-3)

$k_s/D = 0.00015$ (Fig. 5-5)

Assume $f = 0.013$

Then $h_p = 30 + (0.5+1.0+2(0.35)+2(0.2)+0.13(2\text{x}5280/1))V^2/2g$

$$h_p = 30 + 139.9\, V^2/2g$$

$$= 30 + 139.9\, Q^2/(2gA^2); \quad A = \pi/4 = 0.7854\ \text{ft}^2$$

$$h_p = 30 + 3.522Q^2 \text{ (with Q in cfs)}$$

$$= 30 + 1.747 \times 10^{-6}Q^2 \text{ (with Q in gpm)}$$

Plotting the above equation (system curve) on the performance curve for problem 8-12 yields a discharge of 1,500 gpm = 3.34 cfs

$$V = Q/A = 4.26\ \text{ft/sec}$$

check f: $Re = Vd/\nu = 4.26\ (1)\ /\ (1.2 \times 10^{-5}) = 3.55 \times 10^5$

$$f = 0.016$$

With this larger f the system equation becomes:

$$h_p = 30 + 171.6\, V^2/2g$$

$$= 30 + 2.664V^2$$

$$= 30 + 4.323\, Q^2 \text{ (Q in cfs)}$$

Plotting this new system curve, etc. yields Q = <u>1,450 gpm</u>

===

8-34. Assume same system curve as for the solution to problem 8-32:

$h_p = 30 + 4.32\ Q^2$

The parallel pump performance curve is given below:

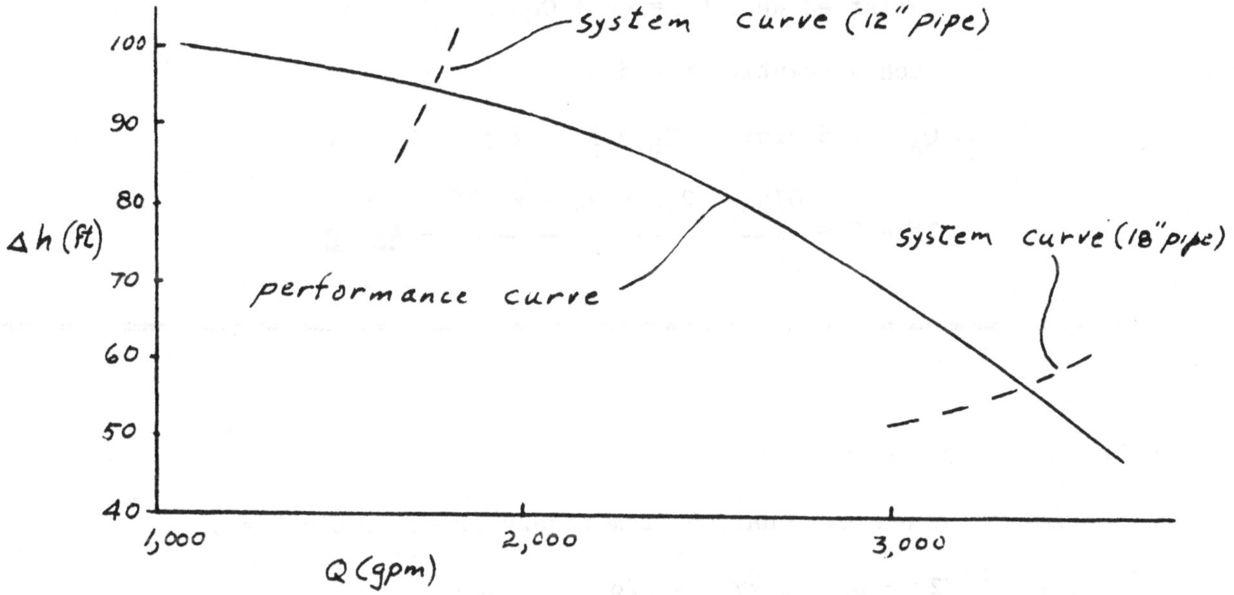

Plotting the system curve on the performance curve yields a solution of $Q \approx$ 1,750 gpm

---

8-35. $k_s/D \approx 0.0001$; assume $f = 0.014$

Then from solution to problem 8-32 the energy equation reduces to

$h_p = 30 + (2.6 + .014\ (2 \times 5,280/1.5))V^2/2g$

$= 30 + 101.2\ V^2/2g$

$= 30 + 101.2\ Q^2/(2gA^2)$; $A = (\pi/4)\ (1.5^2) = 1.767\ ft^2$

$h_p = 30 + 0.503\ Q^2$

Plotting the above equation on the graph of solution for problem 8-33 yields Q = 3,400 gpm

---

8-36. One way to solve this problem is to assume Q_A from which one can get H_{pump}. Then apply the energy Eq. from the upstream reservoir to just downstream of pump--this computation involves H_{LA} and allows one to compute the head downstream of the pump. Then with this head compute Q_B and Q_C. Trial and error solution is satisfied when $Q_A = Q_B + Q_C$.

Such a solution yields

$Q_A \approx 2.6$ cfs; $Q_B = 1.1$ cfs; $Q_C = 1.5$ cfs

Then $P = \frac{Q\gamma h_p}{550} = \frac{2.6 \times 62.4 \times 158}{550} = \underline{\underline{47 \text{ hp}}}$

8-37. Assume T = 10°C

Writing the energy equation from reservoir to turbine jet,

$$p_1/\gamma + V_1^2/2g + z_1 = p_2/\gamma + V_2^2/2g + z_2 + \Sigma h_L$$

$$0 + 0 + 650 = 0 + V_{jet}^2/2g + 0 + (fL/D)(V_{pipe}^2/2g)$$

but from continuity, $V_{pipe}A_{pipe} = V_{jet}A_{jet}$

$$V_{pipe} = V_{jet}(A_{jet}/A_{pipe}) = V_{jet}(0.16) = 0.026\, V_{jet}$$

so, $(V_{jet}^2/2g)(1 + (fL/D)0.026^2) = 650$

$$V_{jet} = [(2 \times 9.81 \times 650)/(1+(0.016(10{,}000)/1)0.026^2)]^{1/2} = 107.3 \text{ m/s}$$

Power $= Q\gamma V_{jet}^2 e = 107.3(\pi/4)(0.16)^2 9{,}810(107.3)^2 0.83/(2 \times 9.81) = \underline{\underline{10.31 \text{ MW}}}$

$V_{bucket} = (1/2)V_{jet} = 53.7$ m/s $= (D/2)\omega$; $D = 53.7 \times 2/(360 \times (\pi/30)) = \underline{\underline{2.85 \text{ m}}}$

8-38. Assume atmospheric pressure head = 10.3 m

Assume $T = 10^\circ C$; $p_v = 1.23\ kN/m^2$; $h_v = p/\gamma = 0.12m$

$n = 35.6$ rps

$Q = 0.27\ m^3/s$

$V = Q/A = 0.27/((\pi/4) \times 0.3^2) = 3.82\ m/s$

$V^2/2g = 0.74\ m$

Assume a head loss coefficient of 1.0 for the inlet and pipe from the inlet to the suction side of the pump. Next, write the energy equation from the reservoir water surface to the inlet side of the pump to determine the pressure on the suction side of the pump.

$p_1/\gamma + z_1 + V^2/2g = p_s/\gamma + z_s + V_s^2/2g + \Sigma h_L$

$10.3 + 100 + 0 = p_s/\gamma + 103 + 0.74 + 1 \times 0.74$

$p_s/\gamma = 10.3 - 3 - 1.48$

$p_s/\gamma = 5.82\ m$

$NPSH = p_s/\gamma - H_V$

$= 5.82 - 0.12$

$= 5.70\ m$

Now compute n_{ss}

$n_{ss} = nQ^{1/2}/(g^{3/4} \times NPSH^{3/4})$

$= 35.6 \times 0.27^{1/2}/(9.81^{3/4} \times (5.70^{3/4})$

$= 35.6 \times 0.52/(5.54 \times 3.69)$

$\underline{= 0.905}$

This value of n_{ss} is above the critical value of 0.494; therefore, **cavitation may occur with this installation.**

8-39. $V_{r_1} = 4/(2\pi \times 1.5 \times 0.3) = 1.415$ m/s; $V_{r_2} = 4/(2\pi \times 1.2 \times 0.3) = 1.768$ m/s;

$\omega = (60/60)2\pi = 2\pi\ s^{-1}$

$\alpha_1 = \text{arc cot}\ ((r_1\omega/V_{r_1}) + \cot\beta_1) = \text{arc cot}((1.5(2\pi)/1.415) + \cot 85°)$

$\alpha_1 = \text{arc cot}(6.66 + 0.0875) = \underline{\underline{8°25'}}$

$V_{tan_1} = r_1\omega + V_{r_1} \cot\beta_1 = 1.5(2\pi) + 1.415(0.0875) = 9.549$

$V_{tan_2} = r_2\omega + V_{r_2} \cot\beta_2 = 1.2(2\pi) + 1.768(-3.732) = 0.940$

$T = \rho Q(r_1V_{tan_1} - r_2V_{tan_2}) = 1{,}000(4)((1.5 \times 9.549) - 1.2 \times 0.940) = \underline{\underline{52{,}780\ \text{N-m}}}$

Power $= T\omega = 52{,}780 \times 2\pi = \underline{\underline{331.6\ \text{kW}}}$

8-40. $\omega = 120/60 \times 2\pi = 4\pi\ s^{-1}$; $V_{r_1} = 113/(2\pi(2.5)0.9) = 7.99$ m/s

$\alpha_1 = \text{arc cot}((r_1\omega/V_{r_1}) + \cot\beta_1) = \text{arc cot}\ ((2.5(4\pi)/7.99) + \cot 45°)$

$= \text{arc cot}\ (3.93 + 1) = \underline{\underline{11°28'}}$

8-41. $V_{r_1} = Q/(2\pi r_1 B) = 126/(2\pi \times 5 \times 1) = 4.01$ m/s

$\omega = 60 \times 2\pi/60 = 2\pi$ rad./s

a) $\alpha_1 = \text{arc cot}((r_1\omega/V_{r_1}) + \cot\beta_1) = \text{arc cot}\ ((5 \times 2\pi/4.01) + 0.577) = \underline{\underline{6.78°}}$

$\alpha_2 = \text{arc tan}(V_{r_2}/(\omega r_2)) = \text{arc tan}((4.01 \times 5/3)/(3 \times 2\pi)) = \text{arc tan}\ 0.355$

$= 15.5°$

b) $V_1 = V_{r_1}/\sin\alpha_1 = 4.01/0.118 = 39.97$ m/s; $V_2 = V_{r_2}/\sin\alpha_2 = 20.0$ m/s

$P = \rho Q\omega(r_1V_1\cos\alpha_1 - r_2V_2\cos\alpha_2)$

$P = 998 \times 126 \times 2\pi(5 \times 39.97 \times \cos 6.78° - 3 \times 20.0 \times \cos 15.5°) = \underline{\underline{111.1\ \text{MW}}}$

c) $\underline{\underline{\text{Increase } \beta_2}}$

8-42. From Fig. 8-36 it appears that an impulse turbine would be the proper turbine for these conditions. The head is relatively high and the discharge is relatively small which would yield a small N_s.

Let $N_s = 5.0$ (N_s for most efficient operation (Fig. 8-34))

Efficiency ≈ 91%

$P = Q\gamma\Delta H/550 = 10 \times 62.4 \times 600 \times 0.91/550 = 620$ horsepower

Compute N from Eq. (8-34):

$$N_s = NP^{1/2}/H^{5/4}$$

or $N = N_s\ H^{5/4}/P^{1/2} = 5\ (600^{5/4})/(620^{1/2}) = 596$ rpm

Now get a better value of N to satisfy the requirement of synchronous speed.

$N = 7{,}200/n$

Assume 12 poles will be used ($n = 12$). Then

$N = 7{,}200/12 = \underline{600\text{ rpm}}$

Now determine D:

Assume $\Phi = 0.45$

then $0.45 = (\pi DN/60)/(2g \times \Delta H)^{1/2}$ (8-28)

where $N = 600$ rpm

$g = 32.2$ ft/sec^2

$H = 600$ ft

Then $D = 0.45\ (2g \times 600)^{1/2}/(\pi(600)/60)$

$D = \underline{2.82\text{ ft}}$

8-43. Referring to Fig. 8-36, it appears that for a head of 200 ft that the appropriate turbine for the given conditions would be a Francis turbine.

Let $N_s \approx 45$ (optimum N_s for Francis turbine (Fig. 8-34))

efficiency ≈ 94% (Fig. 8-34)

8-43. (Continued)

then $P = Q\gamma\Delta H \times eff/550 = 1{,}000 \times 62.4 \times 200 \times 0.94/550$

$= 21{,}329$ hp

$N = N_S H^{5/4}/P^{1/2} = 45\ (200)^{5/4}/(21{,}329^{1/2})$

$= 231.75$ rpm

Choose synchronous speed:

$N = 7{,}200/n$ (let $n = 32$)

$= \underline{225.0 \text{ rpm}}$

Now determine D:

Assume $\Phi = 0.70$

$D = \Phi\ (2g\Delta H)^{1/2}/(\pi N/60)$ (from Eq. 8-28)

$D = 0.70\ (64.4 \times 200)^{1/2}/(\pi\ (225)/60)$

$= \underline{6.74 \text{ ft}}$

8-44. Referring to Fig. 8-36, it appears that for a head of 50 ft that the appropriate type of turbine to use would be a Kaplan turbine. Let $N_S = 140$ (optimum N_S for Kaplan turbine (Fig. 8-34)).

Efficiency ≈ 94% (Fig. 8-34)

$P = Q\gamma\Delta H \times eff./550 = 3{,}000 \times 62.4 \times 50 \times .94/550 = 16{,}000$ hp

$N = N_S\ H^{5/4}/P^{1/2} = 140 \times 50^{5/4}/(16{,}000)^{1/2} = 147.2$ rpm

Choose synchromous speed:

$N = 7{,}200/n$ (let $n = 48$ poles)

Then $N = 7{,}200/48 = 150$ rpm

Now determine D:

Assume $\Phi = 2.0$

$D = \Phi\ (2g\ H)^{1/2}/(\pi N/60)$ (from Eq. 8-28)

$D = 2.0\ (64.4 \times 50)^{1/2}/(\pi \times 150/60) = \underline{14.45 \text{ ft}}$

CHAPTER NINE

9-1. From Fig. 9-2 $Q_p = 310\ cfs/mi^2$

$= 310\ (500) = 155{,}000\ cfs$

$S = 0.0002,\ n = 0.025$

Area: $(2000+y)y$, Wetted Perimeters $P = 2000 + 2(1.414)y$

Hydraulic Radius $= R = A/P = y(2000+y)/[2000+2.828y]$

From Eq. 9-8: $Q = (1.5/n)A\ R^{2/3}S^{1/2}$

$$155000 = (1.5/0.025)[y(2000+y)][y(2000+y)/(2000+2.828y]^{2/3}(.0002)^{1/2}$$

$$182669 = [y(2000+y)][y(2000+y)/(2000+2.828y)]^{2/3}$$

Solve this equation by trial and error as follows:

Assumed y (ft)	R.H.S.	
10	92718	too small
20	294130	too large
15	182185	slightly too small
15.1	184212	slightly too large

Use <u>$y = 15.0$ ft</u>

9-2. $Q = 50.750$ cfs; $S = 0.0045$

Eq. 9-13: $Q = (1.5/n)[A_1R_1^{2/3}+A_2R_2^{2/3}]S^{1/2}$

Let A_1 & R_1 denote the main channel

A_2 & R_2 denote the total overbank channel

$A_1 = (10(200) = 2000\ ft^2$, $P_1 = 200 + 2(5) = 210$

$R_1 = 2000/210 = 9.523$

$R_1^{2/3} = 4.493$

$A_2 = 2(200)5 = 2000\ ft^2$, $P_2 = 200 + 200 + 2(5) = 410$

$R_2 = 2000/410 = 4.878$, $R_2^{2/3} = 2.876$

$n = (1.5/50750)[2000(4.493) + 2000(2.876)](.0045)^{1/2}$

$= 0.436(.0671) = \underline{0.029}$

9-3. Eq. 9-12: $Q = 1.5\ [A_1R_1^{2/3}/n_1+A_2R_2^{2/3}/n_2]S^{1/2}$

$50750 = 1.5\ [2000(4.493)/0.02 + 2000(2.876)/n_2](.0045)^{1/2}$

$504357 = 449300 + 5752/n_2$

$n_2 = 5752/(504357 - 449300)$

$= \underline{0.104}$

9-4. $S = 0.0009$, $Q = 100{,}000$ cfs

From Table 9-4: n(overbank) = 0.06 = n_2
From Table 4-1: n(main channel) = 0.025 = n_1

The solution for y must be made differently for y < 20 and y > 20. Initially we assume y < 20

$A_1 = (500 + y)y$, $P_1 = 500 + 2.828y$

Eq. 9-12: $Q = 100{,}000 = (1.5/0.025)(500+y)y[(500+y)y/(500+2.828y)]^{2/3}(.0009)^{1/2}$

Solve for y by trial and error:

Assumed y (ft)	R.H.S.	
19	67,246	y is too large
16	50,530	y is too small
17	55,890	y is slightly too large
<u>16.9</u>	55,344	OK (y < 20)

9-5. $Q = 45{,}400$, $S = 0.004$, $n_1 = 0.025$, $n_2 = 0.07$

Assume y > 10 but < 18

Main Channel: $A_1 = 102.5(10)+[105+(y-10)/8](y-10)$
$= 1025 + (103.75+0.13y)(y-10)$

$P_1 = 100+(10^2+2.5^2)^{1/2}+[y^2+(y/4)^2]^{1/2}$
$= 100 + 10.3 + 1.03y$
$= 110.3 + 1.03y$

Problem 9-5 Cont'd.

Overbank: $A_2 = 200(y-10)$, $P_2 = 200+y-10 = 190+y$

$$45400/(1.5)(.004)^{1/2} = [1025+(103.75+.13y)(y-10)]$$
$$[(1025+(103.75+.13y)(y-10))/(110.3+1.03y)]^{2/3}/.025$$
$$+\ 200(y-10)[200(y-10)/(190+y)]^{2/3}/0.07$$

$$478558 = [1025+(103.75+.13y)(y-10)][(1025+(103.75+.13y)$$
$$(y-10))/(110.3+1.03y)]^{2/3}/0.025\ +$$
$$200(y-10)[200(y-10)/(190+y)]^{2/3}/0.07$$

Solve for y by trial and error

Assumed y (ft)	R.H.S.	
12	239,543	y is too small
15	373,168	y is too small
17	478,560	OK

Thus for Main Channel: y = 17.0 ft

$$Q(\text{Main}) = (1.5/.025)(.004)^{1/2}[102.5+(103.75+.13y)$$
$$(y-10)][1025+(103.75+.13y)(y-10))/(110.3+1.03y)]^{2/3}$$
$$= 3.794\ (10176) = \underline{38607\ \text{cfs}}$$

Q(Overbank) = 45400-38607 = 6793 cfs

$$(\text{Check})\ Q = (1.5/0.07)(.004)^{1/2}(200(y-10))$$
$$[200(y-10)/(190+y)]^{2/3} = 1.355\ (5007) = \underline{6784\ \ \text{OK}}$$

9-6. $Q = 85,000$, $n_1 = 0.025$, $n_7 = 0.07$, $S = 0.004$

Judging from the results of Prob. 9-5, the flow depth in the main channel will be more than 18 ft.

Main Channel: $A_1 = 1025+[103.75+0.13(18)](18-10)+107(y-18)$
$= 1025+848.7+107(y-18) = 107y - 52.3$

$P_1 = 100+(10^2+2.5^2)^{1/2}+(18^2+18^2/16)^{1/2}$

Overbank: $A_2 = 200(y-10)+100(y-18) = 300y-3800$

$P_2 = 200+(y-10)+100+(y-18) = 272+2y$

$$85000/1.5(.004)^{1/2} = (107y-52.3)[(107y-52.3)/128.9]^{2/3}/0.025$$
$$(300y-3800)[300y-3800)/(272+2y)]^{2/3}/.07$$

$$895979 = (107y-52.3)[(107y-52.3)/128.9]^{2/3}/.025 + (300y-3800)$$
$$[(300y-3800)/(272+2y)]^{2/3}/.07$$

Solve for y by trial and error.

Assumed y (ft)	R.H.S.	
20	650,137	y is too small
25	1,051,030	y is too large
21	723,410	y is too small
22	800,263	y is too small
23	880,560	y is too small
23.1	888,775	y is too small
23.2	897,110	OK

Thus, depth in main channel = <u>23.2 ft</u>

9-7. $Q = 172{,}000$ cfs, $S = 0.0015$

From Table 9-4: n(overbank) = $n_2 = 0.035$
From Table 4-1: n(main) = $n_1 = 0.040$

Assume depth in main channel is greater than 20 feet

Main Channel: $A_1 = 320(20) + 340y = 6400 + 340y$
$P_1 = 300 + 2\,(20^2+20^2)^{1/2} = 356.6$

Overbank: $A_2 = (500+y)y$
$P_2 = (500+2(1.414)y = 500 + 2.828y$

Eq. 9-12: $17200/(1.5)(.0015)^{1/2} = (6400+340y)[(6400+340y)/356.6]^{2/3}/0.040 + (500+y)y[(500+y)y/(500+2.828y)]^{2/3}/0.035$

$2{,}960{,}681 = (6400+340y)[(6400+340y)/356.6]^{2/3}/.040 + (500+y)y[(500+y)y/(500+2.828y)]^{2/3}/.035$

Solve for y by trial and error.

Assumed y (ft)	R.H.S.	
10	2,891,803	y is too small
11	3,135,809	y is too large
10.1	2,915,742	OK

Thus, y = 10.1 feet, height of Levee = 10.1 + 2 = <u>12.1 ft</u>

Mannings Eq. $V = (1.5/n)R^{2/3}\,S^{1/2}$

$V = (1.5/.035)[(500+y)y/(500+2.828y)]^{2/3}(.0015)^{1/2}$
V = <u>7.6 fps</u>

9-8. $N_1 = 0.040$, $n_2 = 0.060$, $S = 0.008$

$A_1 = 215(15)+230(10) = 5525$
$P_1 = 200+2(15^2+15^2)^{1/2} = 200+42.4 = 242.4$
$A_2 = 2[207.5(10)] = 4150$
$P_2 = 2[200+(10^2+15^2)^{1/2}] = 436.1$

$Q = 1.5(.008)^{1/2}[5525(5525/242.4)^{2/3}/.04 + 4150$
$(4150/436.1)^{2/3}/.06] = \underline{190,642\text{ cfs}}$

9-9. $Q(100) = 56000\text{cfs}$, $n_1 = 0.035$, $n_2 = 0.06$,

We must determine S

Eq. 9-13: $Q = (k_1+k_2)S^{12}$
$S^{1/2} = Q/k_1+k_2)$
for $Y = 185 - 160 = 25$
$A_1 = (80+15)20 + 110(5) = 2450\text{ ft}^2$
$P_1 = 80 + 2\ (15^2+20^2)^{1/2} = 130\text{ ft}$
$A_2 = 102.5(5) + 72.5(5) = 875\text{ ft}^2$
$P_2 = 170 + 2(25+25)^{1/2} = 184.1\text{ ft}$

$S^{1/2} =(56000/1.5)[2450(2450/130)^{2/3}/.035+875(875/184.1)^{2/3}/.06]$
$= 37333/536955 = 0.0695$

Now determine width of floodway.

Water Surface Elev. would be 185 + 1 = 186

$\bar{y} = 186 - 160 = 26\text{ ft.}$
$A_1 = (80+15)20 + 110(6) = 2560\text{ ft}^2$
$P_1 = 130\text{ ft.}$
$A_2 = (B+6)6$
$P_2 = B + 2(36+36)^{1/2} = B + 17.0$

Problem 9-9 Cont'd.

Where B = total contracted bottom width of overbank

$$56000/(1.5)(.0695) = 2560(2560/130)^{2/3}/.035 + 6(B+6)[6(B+6)/(B+17)]^{2/3}/.06$$

$$537170 = 533378 + 6(B+6)[6(B+6)/(B+17)]^{2/3}/.06$$

Solve for B by trial and error.

Assumed B (ft)	R.H.S.	
120	572,724	B is too large
100	566,000	B is too large
50	549,785	B is too large
20	540,164	B is too large
10	537,106	OK

Thus, Floodway width = 80 + 15 + 15 + 10 = <u>120 ft</u>

9-10. $Q = 50{,}000$ cfs, $n = 0.030$, $S_0 = 0.0015$

$A = (100+y)y$

$P = 100 + 2.828y$

The solution is to proceed upstream. We will make the solution in 2 steps of 1000 feet each. Smaller steps should be used for greater accuracy. In this case we are computing upstream and the last column of the table (the check) is different than before.

See the following sketch. For this case we use $(E_0 + S_f\Delta x) - (E + S_0\Delta x)$ since this quantity must be zero when we have the correct value of E. The following table gives the solution:

Table for Problem 9-10

Q = 50,000 cfs

Sec	Y	A	P	R	$R^{2/3}$	n		V	$V^2/2g$	E	S_f	Avg. S_f	$S_f \Delta x$	$(E_0 + S_f \Delta x)$ $-(E + S_0 \Delta x)$
0+00	30	3900	184.8	21.10	7.64	.03	1.0	12.82	2.55	32.55	.0011			
-10+00	29	3741	182.0	20.55	7.50	.03	1.0	13.36	2.77	31.72	.0013	.0012	1.20	+0.48
	29.1	3757	182.3	20.61	7.52			13.30	2.75	31.85	.0012	.00115	1.15	+0.35
	29.3	3788	182.9	20.71	7.54			13.19	2.70	32.00	.0012	.00115	1.15	+0.20
	29.5	3820	183.4	20.83	7.57			13.10	2.66	32.16	.0012	.00115	1.15	+0.04
-20+00	29	3741	182.0	20.55	7.50	.03	1.0	13.36	2.77	31.77	.0013	.00125	1.25	+0.14
	29.2	3773	182.6	20.66	7.53			13.25	2.72	31.92	.0012	.0012	1.20	-0.06

Problem 9-10 Cont'd.

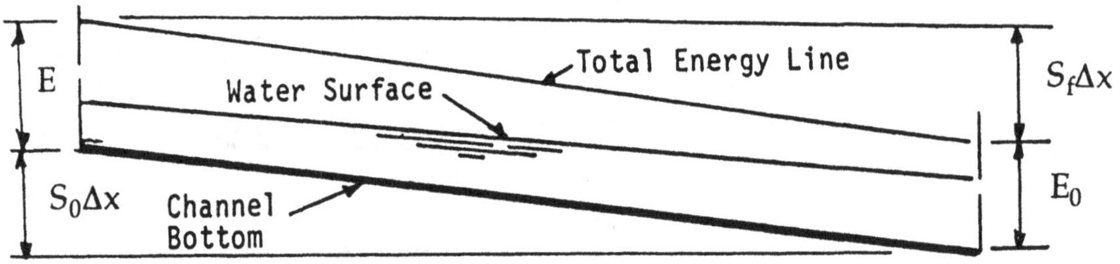

The table shows y = 29.5 @ X = -1000
y = 29.2 @ X = -2000

Thus, W.S. Elev. = 30 @ X = 0
= -1.5 + 29.5 = <u>28.0 @ X = -1000</u>
= -3.0 + 29.2 = <u>26.2 @ X = -2000</u>

9-11. Q = 65,000 cfs, n_1 = 0.030, n_2 = 0.050
S = 0.005
@ x = 0, y = 14 ft.

Write equations for A, P, & R for main channel and overbank area.

Main: $A_1 = 200y$
$P_1 = 200 + 5 + 5 = 210$
$A_2 = 400(y-5)$
$P_2 = 400 + 2(y-5) = 390 + 2y$

Use these equations to complete R, A, and P values in the following Table. We use a single step of 4000 feet. A slightly different depth will be obtained if several smaller steps are used. Thus, y(4000) = <u>11.0 ft</u>

Table for Prob. 9-11.

Sec	Sub Area	Y (ft)	A (ft)	P (ft)	R (ft)	$R^{2/3}$	n	K	K^3/A^2	α	V (fps)	$\alpha V^2/2g$	E	S_f	Avg S_f	(17)	(18)
0+00	1	14	2800	210	13.3	5.61	.03	785400	6.180								
	2		3600	418	8.6	4.19	.05	452520	0.715	1.489	10.15	2.38	16.38	.0028			
			6400					1237970	6.895								
40+00	1	15	3000	210	14.3	5.89	.03	883500	7.663								
	2		4000	420	9.5	4.49	.05	538800	0.978	1.471	9.28	1.97	16.97	0.0021	.00245	9.8	+9.61
			7000					1422300	8.641								
	1	10	2000	210	9.52	4.49	.03	449000	2.263								
	2		2000	400	5.00	2.92	.05	175200	0.134	1.577	16.25	6.47	16.47	.0108	.0068	27.20	-7.29
			4000					624200	2.397								
	1	11	2200	210	10.48	4.79	.03	526900	3.022								
	2		2400	412	5.83	3.24	.05	233280	.220	1.562	14.13	4.84	15.84	.0073	.0051	20.4	+0.14
			4600					760180	3.242								
	1	10.9	2180	210	10.38	4.76	.03	518840	2.929								
	2		2360	411.8	5.73	3.20	.05	226560	.209	1.567	14.32	4.99	15.89	.0076	.0052	20.8	-0.31
			4540					745400	3.148								

Note: Column 17 is 4000 S_f

Column 18 is $(E_0 + S_f\Delta x) - (E + S_0\Delta x)$

9-12. N = 10 years

Initial Cost = $12,000

Probable Annual Damages = 12000(0.01)	= $ 120.00
Annual Maintenance	= $ 600.00
Neglecting Interest, Cost of Installation = 12000/10 annually	= $1200.00
Total Annual Cost	$1920.00

9-13. $P_{50} = 0.02$, $P_{25} = 0.04$

Prob. of Q being between Q_{50} and Q_{25}	= 0.02
Annual Cost of Installation = 75000/25	= $3000.00/yr.
Probable Annual Damages = 0.02(75000/2)	= $ 750.00/yr.
Total Annual Cost	= $3750.00/yr.

We have neglected interest and maintenance.

9-14. $P_{50} = 0.02$, $P_{10} = 0.10$

From Figure: $Q_{50} = 7700$ cfs, $Q_{10} = 4200$

Set up the following Table Damage: 200,000 (Q-4200)/(7700-4200)

Q (cfs)	Damage ($)	PEXC	Prob. in Range	Avg. Damage in Rouge ($)	Prob. Damage ($)
4200	0	0.10			
			0.039	22,857	891
5000	45714	0.061			
			0.024	74,286	1,782
6000	102,857	0.037			
			0.012	131,428	1,577
7000	160,000	0.025			
			0.005	180,000	900
7700	200,000	0.02			
		Average Probable Annual Damage:			$5,150

9-15.

Flood Stage (ft)	Damage ($)	P_{EXC}	Ave. Damage in Range ($)	Prob. in Range	Prob. Damages in Range ($)	Accum. Annual Damages ($)	Annual Cost of Const. ($)
35	0	.495				0	0
			50,000	0.080	4,000		
36	100,000	.415				4,000	4,000
			150,000	0.080	12,000		
37	200,000	.335				16,000	9,000
			250,000	0.080	20,000		
38	300,000	.255				36,000	14,000
			375,000	0.070	26,250		
39	450,000	.185				62,250	22,000
			525,000	0.050	26,250		
40	600,000	.135				88,500	31,000
			700,000	0.044	30,800		
41	800,000	.091				119,300	41,000
			900,000	0.0332	29,880		
42	1,000,000	.0578				149,180	51,000
			1,125,000	0.0325	36,562		
44	1,250,000	.0253				185,742	71,000
			1,400,000	0.0158	22,120		
46	1,550,000	.0095				207,862	91,000
			1,750,000	.0065	11,375		
48	1,950,000	.0030				319,237	121,000
			2,200,000	.0020	4,400		
50	2.450,000	.0010				223,637	151,000
			2,700,000	.0005	1,350		
52	3,950,000	.0005				224,987	191,000
			5,050,000	.0005	2,525		
>52	6,150,000	0				227,512	

Yes, it is feasible since benefits > cost at all heights.

For maximum benefit it should be constructed at least:

a) to Flood Stage 52 since benefits > cost to that point

b) to Flood Stage 44 since above that stage benefits increase slower than costs.

9-16. From Table Q_D = 30 cfs, P_{10} = 0.10 = 10%

Q (cfs)	P_{EXC}	QAVG (cfs)	D in Range ($)	Avg. Damage in Range ($)	Prob. Damage in Range ($)
30	0.10	32.5			
			.02	92.50	1.85
35	0.08	37.5			
			.02	127.50	2.55
40	0.06	42.5			
			.01	162.50	1.63
45	0.05	47.5			
			.01	197.50	1.98
50	0.04	52.5			
			.005	232.50	1.16
55	0.035	57.5			
			.010	267.50	2.68
60	0.025	62.5			
			.005	302.50	1.51
65	0.020	67.5			
			.005	337.50	1.69
70	0.015	72.5			
			.003	372.50	1.18
75	0.012	77.5			
			.002	407.50	0.82
100	0.010	125.0			
			.010	740.00	7.40
150	0				

Probable Annual Damage = $24.45

9-17. Use the given hydrograph to obtain inflow rates for the following table. Use a $\Delta t = 4$ hours since $2S/\Delta t + O$ values are available from the table for Ex. 9-4. Note that 8000 cfs outflow = 8,000(1.08/7) = 660 Acre feet/hour.

Time (hrs)	Inflow (cfs)	(AF/hr)	S_1 (AF)	O_1 (AF/hr)	O_1 (AF/hr)	O_2 (AF/hr)	Reservoir Stage (ft)
0	700		1969500			905411	1049.5
4	2000	223		58	985295	985518	
8	3300	437		165	985188	985625	
12	4650	656		272	985081	985737	
16	6000	879		384	984969	985848	
20	10250	1341		495	984858	986199	
24	14500	2042		660	984879	986921	
28	16750	2578		660	985601	988179	
32	19000	2949		660	986859	989808	
36	18250	3073		660	988488	991561	
40	17500	2949		660	990241	993190	
44	16750	2826		660	991870	994696	
48	16000	2702		660	993376	996078	
52	15000	2558		660	994758	997316	
56	14000	2392		660	995996	998388	
60	13000	2228		660	997068	999296	
64	12000	2062		660	997976	1000038	
68	10000	1815		660	998718	1000533	
72	8000	1485		660	999213	1000698	
76	6000	1155		660	999378	1000533	
80	4000	825					
84	3100	586					

Thus, the reservoir reaches its maximum elevaton at the end of 72 hours. Releases are kept to the rate of inflow until 20 hours when the inflow rate begins to exceed 8000 cfs. According to Table a for example 9-4., the maximum reservoir elevation would be between 1052 and 1054 which is only slightly above the lower limit of flood control storage.

9-18. At Elevation 1088, storage = 2,420,000 AF
At Elevation 1049.5 storage = 985,411 AF
Difference = 1,434,589 AF

Thus, if the reservoir were at El. 1049.5 when flood inflow began a maximum inflow volume of 1,434,589 AF could be stored before it was necessary to exceed 8000 cfs outflow.

9-19.

Dam Height (ft)	P_{EXC}	Annual Damages ($)	Prob. Annual Benefit ($)	Annual Cost of Construction ($)	Incr. Annual Benefits ($)	Incr. Annual Costs ($)
100	0.01	6,000,000	60,000	60,000		
150	0.005	15,000,000	75,000	70,000	15,000	10,000
200	0.002	76,000,000	152,000	140,000	77,000	70,000
250	0.001	160,000,000	160,000	400,000	8,000	260,000
300	0.0005	400,000,000	1,200,000	1,500,000	40,000	1,100,000

Thus, if the dam is built to a height of 200 ft probable annual benefits will exceed annual costs demonstrating that the project is economically feasible.

Incremental annual benefits exceed incremental annual costs for dam heights up to 200. Therefore, a dam height of 200 is also the most economical height.

10-1 i)Bisection method
Newton-Raphson method

ii)For this problem I would select the Bisection method. The Bisection method might take longer, but you don't have to find the derivative of the function (in case the derivative might be zero)

iii)Bisection method

a) $F(y_2) = Q - \frac{1}{n} A R^{2/3} S_0^{1/2} = 0$

b) Choose y_p such that $F(y_p) > 0$
Choose y_n such that $F(y_n) < 0$

c) $y_e = \frac{1}{2}(y_p + y_n)$

d) Determine $F(y_e)$
if $|F(y_e)| < \varepsilon$ where ε is a specified tolerance, then y_e = root, if not let

$y_e = y_n$ if function is negative or let $y_e = y_p$ if function is positive.

e) Repeat steps c) and d) until you get under specified tolerance.

10-2 i)Bisection method
Newton-Raphson method

ii) $F_X = H_{PC} - H_{PS}$

$H_{PS} = 100 + 0.1Q^2$

$H_{PC} = 120 - 0.15Q^{1.8}$

$F(x) = 20 - 0.1Q^2 - 0.15Q^{1.8}$

Using Newton-Raphson method

$x_n = x_0 - F(x_0) / F'(x_0)$

$F'(x) = -1.8(0.15)Q^{0.8} - 2(0.1)Q^1$

$$F'(x) = -0.27Q^{0.8} - 0.2Q$$

Guess Q = 20 cfs

$$Q_n = 20 - \frac{20-(0.15)(20)^{1.8} - 0.1(20)^2}{-0.27(20)^{0.8} - 0.2(20)}$$

$$Q_n = 12.4$$

$$Q_n = 12.4 - \frac{20-(0.15)(12.4)^{1.8} - 0.1(12.4)^2}{0.27(12.4)^{0.8} - 0.2(12.4)}$$

$$Q_n = 10.33$$

$$Q_n = 10.33 - \frac{20-0.15(10.33)^{1.8} - 0.1(10.33)^2}{-0.27(10.33)^{0.8} - 0.2(10.33)}$$

$$\underline{Q_n = 10.15cfs}$$

10-3 Parallel pumps : same H, add Q's

$$H_A = H_B$$
$$300 - 0.1Q_A{}^2 = 335 - 0.2Q_B{}^2$$
$$0.2Q_B{}^2 - 0.1Q_A{}^2 = 35$$
$$Q_B + Q_A = 25$$
$$Q_A = 25 - Q_B$$
$$0.2Q_B{}^2 - 0.1(25 - Q_B)^2 = 35$$
$$0.2Q_B{}^2 - 0.1(625 - 50Q_B + Q_B{}^2) = 35$$
$$0.2Q_B{}^2 - 62.5 + 5Q_B - 0.1Q_B{}^2 = 35$$
$$0.1Q_B{}^2 + 5Q_B - 62.5 = 35$$
$$Q_B{}^2 + 50Q_B - 975 = 0$$
$$Q_B = \frac{-50 \pm \sqrt{50^2 - 4(1)(-975)}}{2(1)}$$
$$Q_B = 15, -65$$
$$\rightarrow Q_B = 15cfs$$
$$Q_A = 25 - 15 = 10cfs$$

check:

$$0.2(15)^2 - 0.1(10)^2 = 35 \quad \text{o.k}$$

10-4 i)Bisection method
Newton-Raphson method

ii)For this problem I would select the Bisection method. The Bisection method might take longer, but you don't have to find the derivative of the function (in case the derivative might be zero)

iii)Bisection method

a)

$$F(y_2) = y_2 + z_2 + (1+k)\frac{Q^2}{2gA_2^{\,2}} - z_1 - y_1 - \frac{Q^2}{2gA_1^{\,2}} = 0$$

b) Choose y_p such that $F(y_p)\rangle 0$
Choose y_n such that $F(y_n)\langle 0$

c) $y_e = \frac{1}{2}(y_p + y_n)$

d) Determine $F(y_e)$
if $|F(y_e)|\langle \varepsilon$ where ε is a specified tolerance, then y_e = root, if not let

$y_e = y_n$ if function is negative or let $y_e = y_p$ if function is positive.

e) Repeat steps c) and d) until you get under specified tolerance.

10-5 i)Nonlinear algebraic equation

ii)Linear ordinary differential equation

iii)Nonlinear partial differential equation

Computational steps for the solution of $kQ|Q| = H_0 - Q^{2.1}$ using Newton-Raphson method

a) Determine an expression for F(y)

$$F(y) = kQ|Q| + Q^{2.1} - H_0 = 0$$

b) Determine the derivative

$$F'(y) = 2kQ + 2.1Q^{1.1}$$

c) Guess a value for y and determine y_n from the following equation:

d) $y_n = y - F(y) / F'(y)$

e) If $|(y_n - y)| \langle \varepsilon$ where ε is a specified tolerance, then x_n is the root

otherwise, set $y = y_n$

f) Repeat d) and e) until a specified tolerance is achieved

10-6 i)Steady uniform:

$$\frac{\partial y}{\partial t} = 0$$

$$\frac{\partial y}{\partial x} = 0$$

ii)Steady non-uniform:

$$\frac{\partial y}{\partial t} = 0$$

$$\frac{\partial y}{\partial x} \neq 0$$

iii)Unsteady uniform

$$\frac{\partial y}{\partial t} \neq 0$$

$$\frac{\partial y}{\partial x} = 0$$

iv)Unsteady non-uniform

$$\frac{\partial y}{\partial t} \neq 0$$
$$\frac{\partial y}{\partial x} \neq 0$$

10-7 $$E = y + \frac{q^2}{2gy^2}$$

$$1.5 = y + \frac{1^2}{2(9.81)y^2}$$

$$F(y) = y + \frac{1}{19.62y^2} - 1.5 = 0$$

$$F'(y) = 1 - \frac{2}{19.62y^3}$$

$$y_n = y_0 - F(y_0)/F'(y_0)$$

put $y_0 = 1$

$$y_n = 1 - \frac{1 + 1/19.62(1)^2 - 1.5}{1 - 2/19.62(1)^3} = 1.5$$

$$y_n = 1.5 - \frac{1.5 + 1/19.62(1.5)^2 - 1.5}{1 - 2/19.62(1.5)^3}$$

$$\underline{y_n = 1.477m}$$

10-8 By Taylor's series

$$F(x_0 + \Delta x, y_0 + \Delta y, z_0 + \Delta z) = F(x_0, y_0, z_0) + \frac{\partial F}{\partial x}\Delta x + \frac{\partial F}{\partial y}\Delta y + \frac{\partial F}{\partial z}\Delta z + 0(2) = 0 --- (1)$$

$$G(x_0 + \Delta x, y_0 + \Delta y, z_0 + \Delta z) = G(x_0, y_0, z_0) + \frac{\partial G}{\partial x}\Delta x + \frac{\partial G}{\partial y}\Delta y + \frac{\partial G}{\partial z}\Delta z + 0(2) = 0 --- (2)$$

$$H(x_0+\Delta x, y_0+\Delta y, z_0+\Delta z) = H(x_0,y_0,z_0) + \frac{\partial H}{\partial x}\Delta x + \frac{\partial H}{\partial y}\Delta y + \frac{\partial H}{\partial z}\Delta z + 0(2) = 0 --- (3)$$

Solving equations (1),(2), and (3) for Δx, Δy, and Δz we obtain:

$$\Delta x = \{-F(x_0,y_0,z_0)\frac{\partial F}{\partial y}\frac{\partial H}{\partial y}\frac{\partial G}{\partial z} + F(x_0,y_0,z_0)\frac{\partial G}{\partial y}\frac{\partial H}{\partial y}\frac{\partial G}{\partial z} + G(x_0,y_0,z_0)\frac{\partial F}{\partial y}\frac{\partial G}{\partial y}\frac{\partial H}{\partial z} - G(x_0,y_0,z_0)\frac{\partial F}{\partial y}\frac{\partial H}{\partial y}\frac{\partial G}{\partial z}$$
$$-H(x_0,y_0,z_0)\frac{\partial F}{\partial y}\frac{\partial G}{\partial y}\frac{\partial G}{\partial z} + G(x_0,y_0,z_0)\frac{\partial F}{\partial y}\frac{\partial H}{\partial y}\frac{\partial G}{\partial z} + H(x_0,y_0,z_0)\frac{\partial G}{\partial y}\frac{\partial G}{\partial y}\frac{\partial F}{\partial z} - G(x_0,y_0,z_0)\frac{\partial G}{\partial y}\frac{\partial H}{\partial y}\frac{\partial F}{\partial z}.\}/$$
$$(\frac{\partial G}{\partial y}\frac{\partial F}{\partial x}\frac{\partial G}{\partial y}\frac{\partial H}{\partial z} - \frac{\partial G}{\partial y}\frac{\partial F}{\partial x}\frac{\partial H}{\partial y}\frac{\partial G}{\partial z} - \frac{\partial F}{\partial y}\frac{\partial G}{\partial x}\frac{\partial G}{\partial y}\frac{\partial H}{\partial z} + \frac{\partial F}{\partial y}\frac{\partial G}{\partial x}\frac{\partial H}{\partial y}\frac{\partial G}{\partial z} + \frac{\partial F}{\partial y}\frac{\partial H}{\partial x}\frac{\partial G}{\partial y}\frac{\partial G}{\partial z} - \frac{\partial F}{\partial y}\frac{\partial G}{\partial x}\frac{\partial H}{\partial y}\frac{\partial G}{\partial z} - \frac{\partial G}{\partial y}\frac{\partial H}{\partial x}\frac{\partial G}{\partial y}\frac{\partial F}{\partial z} + \frac{\partial G}{\partial y}\frac{\partial G}{\partial x}\frac{\partial H}{\partial y}\frac{\partial F}{\partial z})$$

$$\Delta y = \{-G(x_0,y_0,z_0)\frac{\partial F}{\partial x}\frac{\partial F}{\partial x}\frac{\partial H}{\partial z} + G(x_0,y_0,z_0)\frac{\partial F}{\partial x}\frac{\partial H}{\partial x}\frac{\partial F}{\partial z} + F(x_0,y_0,z_0)\frac{\partial G}{\partial x}\frac{\partial F}{\partial x}\frac{\partial H}{\partial z} - F(x_0,y_0,z_0)\frac{\partial F}{\partial x}\frac{\partial H}{\partial x}\frac{\partial G}{\partial z}$$
$$-H(x_0,y_0,z_0)\frac{\partial F}{\partial z}\frac{\partial G}{\partial x}\frac{\partial F}{\partial x} + F(x_0,y_0,z_0)\frac{\partial F}{\partial z}\frac{\partial G}{\partial x}\frac{\partial H}{\partial x} + H(x_0,y_0,z_0)\frac{\partial F}{\partial x}\frac{\partial G}{\partial z}\frac{\partial F}{\partial x} - F(x_0,y_0,z_0)\frac{\partial F}{\partial x}\frac{\partial G}{\partial z}\frac{\partial H}{\partial x}.\}/$$
$$(\frac{\partial G}{\partial y}\frac{\partial F}{\partial x}\frac{\partial F}{\partial x}\frac{\partial H}{\partial z} - \frac{\partial G}{\partial y}\frac{\partial F}{\partial x}\frac{\partial F}{\partial z}\frac{\partial H}{\partial x} - \frac{\partial F}{\partial y}\frac{\partial G}{\partial x}\frac{\partial H}{\partial z}\frac{\partial F}{\partial x} + \frac{\partial F}{\partial y}\frac{\partial G}{\partial x}\frac{\partial H}{\partial x}\frac{\partial F}{\partial z} - \frac{\partial F}{\partial y}\frac{\partial H}{\partial x}\frac{\partial G}{\partial x}\frac{\partial F}{\partial z} + \frac{\partial H}{\partial y}\frac{\partial F}{\partial x}\frac{\partial G}{\partial x}\frac{\partial F}{\partial z} + \frac{\partial F}{\partial y}\frac{\partial H}{\partial x}\frac{\partial G}{\partial x}\frac{\partial G}{\partial z} - \frac{\partial H}{\partial y}\frac{\partial F}{\partial x}\frac{\partial G}{\partial x}\frac{\partial G}{\partial z})$$

$$\Delta z = \{-H(x_0,y_0,z_0)\frac{\partial F}{\partial x}\frac{\partial G}{\partial y}\frac{\partial F}{\partial x} + H(x_0,y_0,z_0)\frac{\partial F}{\partial x}\frac{\partial F}{\partial y}\frac{\partial G}{\partial x} + F(x_0,y_0,z_0)\frac{\partial H}{\partial x}\frac{\partial G}{\partial y}\frac{\partial F}{\partial x} - F(x_0,y_0,z_0)\frac{\partial H}{\partial x}\frac{\partial F}{\partial y}\frac{\partial G}{\partial x}$$
$$-G(x_0,y_0,z_0)\frac{\partial H}{\partial x}\frac{\partial F}{\partial y}\frac{\partial F}{\partial x} + F(x_0,y_0,z_0)\frac{\partial H}{\partial x}\frac{\partial F}{\partial y}\frac{\partial G}{\partial x} + G(x_0,y_0,z_0)\frac{\partial F}{\partial x}\frac{\partial H}{\partial y}\frac{\partial F}{\partial x} - F(x_0,y_0,z_0)\frac{\partial F}{\partial x}\frac{\partial H}{\partial y}\frac{\partial G}{\partial x}.\}/$$
$$(\frac{\partial G}{\partial y}\frac{\partial F}{\partial x}\frac{\partial F}{\partial x}\frac{\partial H}{\partial z} - \frac{\partial F}{\partial y}\frac{\partial G}{\partial x}\frac{\partial F}{\partial x}\frac{\partial H}{\partial z} - \frac{\partial G}{\partial y}\frac{\partial H}{\partial x}\frac{\partial F}{\partial x}\frac{\partial F}{\partial z} + \frac{\partial F}{\partial y}\frac{\partial H}{\partial x}\frac{\partial G}{\partial x}\frac{\partial F}{\partial z} - \frac{\partial F}{\partial y}\frac{\partial H}{\partial x}\frac{\partial G}{\partial x}\frac{\partial F}{\partial z} + \frac{\partial F}{\partial y}\frac{\partial H}{\partial x}\frac{\partial F}{\partial x}\frac{\partial G}{\partial z} - \frac{\partial H}{\partial y}\frac{\partial G}{\partial x}\frac{\partial F}{\partial x}\frac{\partial F}{\partial z} - \frac{\partial H}{\partial y}\frac{\partial F}{\partial x}\frac{\partial F}{\partial x}\frac{\partial G}{\partial z})$$

10-9 i)Bisection method
Newton-Raphson method

ii) Using Newton-Raphson method

computation steps:

a) Determine an expression for F(x)

$$F_X = H_{PC} - H_{PS}$$

$$H_{PS} = 101 + 0.5Q^2$$

$$H_{PC} = 108 - 0.7Q^{1.8}$$

$$F(x) = 7 - 0.5Q^2 - 0.7Q^{1.8}$$

b) Determine the derivative

$$F'(x) = -1.8(0.7)Q^{0.8} - 2(0.5)Q^{1}$$

$$F'(x) = -1.26Q^{0.8} - Q$$

c) Guess a value for x_0 and determine x_n from the following equation:

d) $x_n = x_0 - F(x_0) / F'(x_0)$

e) If $|(x_n - x_0)| \langle \varepsilon$ where ε is a specified tolerance, then x_n is the root

otherwise, set $x_0 = x_n$

f) Repeat d) and e) until a specified tolerance is achieved

10-10 Computation steps using Newton-Raphson Method

a) Determine an expression for F(y)

$$F(y) = A^3 - \frac{Q^2}{g}B = 0$$

b) Determine the derivative

$$F'(y) = 3A^2\frac{dA}{dy} - \frac{Q^2}{g}\frac{dB}{dy} \quad \text{but} \quad \frac{dA}{dy} = B$$

$$F'(y) = 3A^2B - \frac{Q^2}{g}\frac{dB}{dy}$$

c) Guess a value for y and determine y_n from the following equation:

d) $y_n = y - F(y) / F'(y)$

e) If $|(y_n - y)| \langle \varepsilon$ where ε is a specified tolerance, then x_n is the root

otherwise, set $y = y_n$

f) Repeat d) and e) until a specified tolerance is achieved

10-11

$$E_1 = H_1 = y_1 + \frac{v_1^2}{2g}$$

$$v_1 = \frac{600}{30*10} = 2m/s$$

$$H_1 = 10 + \frac{4}{2*9.81} = 10.204m$$

$$E_2 = H_2 = 0.2 + y_2 + \frac{v_2^2}{2g}$$

$$H_2 = H_1 - 0.2\frac{v_2^2}{2g}$$

$$0.2 + y_2 + \frac{v_2^2}{2g} = H_1 - 0.2\frac{v_2^2}{2g} = 10.204 - 0.2\frac{v_2^2}{2g}$$

$$y_2 + 1.2\frac{v_2^2}{2g} = 10.004$$

$$10.004 = y_2 + 1.2\frac{Q^2}{2g(25*y_2)^2} = y_2 + \frac{35.229}{y_2^2}$$

Or

$$y_2^3 - 10.004y_2^2 + 35.229 = 0$$

using Newton-Raphson method

$$F(y_2) = y_2^3 - 10.004y_2^2 + 35.229 = 0$$

$$F'(y_2) = 3y_2^2 - 20.008y_2$$

$$(y_2)_n = y_2 - \frac{F(y_2)}{F'(y_2)} = y_2 - \frac{y_2^3 - 10.004y_2^2 + 35.229}{3y_2^2 - 20.008y_2}$$

start with $y_2 = 10m \rightarrow (y_2)_n = 9.651$

$y_2 = 9.651 \rightarrow (y_2)_n = 9.6238$

$y_2 = 9.6238 \rightarrow (y_2)_n = 9.6236$

$\underline{y_2 = 9.624m}$

10-12

Same as Fig. 10.2

10-13

```
C
C     RESERVOIR ROUTING USING PREDICTOR-CORRECTOR METHOD
C               WITH PARABOLIC INTERPOLATION
C
C     ************** NOTATION ****************
C
C     AR(I) = SURFACE AREA OF RESERVOIR AT LEVEL ELAR(I);
C     DT = ROUTING INTERVAL;
C     NAR = NUMBER OF POINTS ON THE STAGE VS. AREA CURVE;
C     NQO = NUMBER OF POINTS ON THE STAGE VS. OUTFLOW CURVE;
C     NT = NUMBER OF POINTS ON THE INFLOW VS TIME CURVE;
C     QIN(I) = INFLOW AT TIME(I);
C     QO(I) = OUTFLOW AT WATER LEVEL ELQO(I);
C     QO1 = OUTFLOW AT TIME T = 0;
C     Z = RESERVOIR LEVEL ABOVE DATUM;
C     Z1 = RESERVOIR LEVEL AT TIME T = 0;
C     TSTOP = TIME UPTO WHICH ROUTING IS TO BE COMPUTED.
C
      DIMENSION AR(50),ELAR(50),ELQO(50),QO(50),TIME(50),QIN(50)
C
C     INITIAL CONDITIONS
C
      READ(5,*) Z1,QO1,DT,TSTOP
      WRITE(6,20) Z1,QO1,DT,TSTOP
20    FORMAT(//5X,'INITIAL RESERVOIR WATER LEVEL =',F7.2,' M'/
     1 5X,'INITIAL OUTFLOW =',F5.1,' M3/S'/
     2 5X,'ROUTING INTERVAL =',F6.1,' S'/
     3 5X,'TIME UPTO WHICH ROUTING IS TO BE DONE =',F8.1,' S'/)
C
C     STAGE VS. RESERVOIR-SURFACE AREA CURVE
C
      READ (5,*) NAR,(ELAR(I),AR(I),I=1,NAR)
      WRITE(6,30)
30    FORMAT(15X,'STAGE',5X,'RESERVOIR SURFACE AREA'/
     1  16X,'(M)',13X,'(SQ. M)')
      WRITE(6,40) (ELAR(I),AR(I),I=1,NAR)
40    FORMAT(9X,F10.1,10X,F10.1)
C
C     STAGE-OUTFLOW CURVE
C
      READ (5,*) NQO,DQO,(ELQO(I),QO(I),I=1,NQO)
      WRITE(6,60)
60    FORMAT(/15X,'STAGE',9X,'  OUTFLOW'/
     1  16X,'(M)',12X,'(M3/S)')
      WRITE(6,70) (ELQO(I),QO(I),I=1,NQO)
70    FORMAT(9X,F10.1,8X,F10.1)
C
C     INFLOW HYDROGRAPH
C
      READ(5,*) NT, (TIME(I),QIN(I), I=1,NT)
      WRITE(6,90)
90    FORMAT(/14X,'TIME',12X,'INFLOW'/
     1  14X,'(S)',13X,'(M3/S)')
      WRITE(6,100) (TIME(I),QIN(I),I=1,NT)
100   FORMAT(8X,F10.1,8X,F10.2)
```

```
      T = 0.
      WRITE(6,125)
125   FORMAT(//13X,'TIME',11X,'INFLOW',4X,'RESERVOIR LEVEL',2X,
     1   'OUTFLOW'/13X,'(S)',12X,'(M3/S)',9X,'(M)',11X,'M3/S)')
      WRITE(6,240) T,QIN(1).Z1,QO1
C
C      PREDICTOR PART
C
135   T = T+DT
      DO 140 I=1,NT
      IF (T.LT.TIME(I)) GO TO 150
140   CONTINUE
150   I1=I-1
      QIP=QIN(I1)+(T-TIME(I1))/(TIME(I)-TIME(I1))*(QIN(I)-QIN(I1))
      DO 160 I=1,NQO
      IF (Z1.LT.ELQO(I)) GO TO 170
160   CONTINUE
170   I1=I-1
      CALL PARAB(I1,Z1,ELQO,QO,QOP,DQO)
      DO 180 I=1,NAR
      IF (Z1.LT.ELAR(I)) GO TO 190
180   CONTINUE
190   I1=I-1
      ARP=AR(I1)+(Z1-ELAR(I1))/(ELAR(I)-ELAR(I1))*(AR(I)-AR(I1))
      DZP=(QIP-QOP)/ARP
      ZP=Z1+DZP*DT
C
C     CORRECTOR PART
C
      DO 200 I=1,NQO
      IF (ZP.LT.ELQO(I)) GO TO 210
200   CONTINUE
210   I1=I-1
      CALL PARAB(I1,ZP,ELQO,QO,QOC,DQO)
      DO 220 I=1,NAR
      IF (ZP.LT.ELAR(I)) GO TO 230
220   CONTINUE
230   I1=I-1
      ARC=AR(I1)+(ZP-ELAR(I1))/(ELAR(I)-ELAR(I1))*(AR(I)-AR(I1))
      DZC=(QIP-QOC)/ARC
      Z2=Z1+0.5*DT*(DZC+DZP)
      DO 232 I=1,NQO
      IF (Z2.LT.ELQO(I)) GO TO 235
232   CONTINUE
235   I1=I-1
      Q2=QO(I1)+(Z2-ELQO(I1))/(ELQO(I)-ELQO(I1))*(QO(I)-QO(I1))
      WRITE(6,240) T,QIP,Z2,Q2
240   FORMAT(9X,F10.2,5X,F10.2,2X,F10.2,5X,F10.2)
      IF (T.GT.TSTOP) GO TO 250
      Z1=Z2
      GO TO 135
250   STOP
      END
      SUBROUTINE PARAB(I1,ZZ,ELQO,QO,QOZ,DQO)
      DIMENSION ELQO(50),QO(50)


       R = (ZZ-ELQO(I1))/DQO
       IF(I1.EQ.1) R=R-1.0
       IF(I1.LT.2) I1=2
       QOZ = QO(I1)+0.5*R*(QO(I1+1)-QO(I1-1)+R*(QO(I1+1)+QO(I1-1)
     *-2.0*QO(I1)))
       RETURN
       END
```

INITIAL RESERVOIR WATER LEVEL = 0.00 M
INITIAL OUTFLOW = 0.0 M3/S
ROUTING INTERVAL = 900.0 S
TIME UPTO WHICH ROUTING IS TO BE DONE = 10800.0 S

STAGE (M)	RESERVOIR SURFACE AREA (SQ. M)
0.0	176400.0
1.0	179800.0
2.0	183200.0
3.0	186600.0
4.0	190100.0
5.0	193600.0
6.0	197100.0
7.0	200700.0
8.0	204300.0
9.0	207900.0
10.0	211600.0

STAGE (M)	OUTFLOW (M3/S)
0.0	0.0
1.0	20.0
2.0	48.0
3.0	100.0
4.0	160.0
5.0	210.0
6.0	260.0
7.0	300.0
8.0	330.0
9.0	360.0
10.0	380.0

TIME (S)	INFLOW (M3/S)
0.0	0.00
900.0	5.00
1800.0	16.00
2700.0	39.00
3600.0	104.00
4500.0	322.00
5400.0	555.00
6300.0	722.00
7200.0	626.00
8100.0	481.00
9000.0	318.00
9900.0	162.00

NUMBERS IN BRACKETS REFER TO THE RESULTS WITH LINEAR INTERPOLATION

TIME (S)	INFLOW (M3/S)	RESERVOIR LEVEL (M)	OUTFLOW M3/S)
0.00	0.00	0.00 (0.00)	0.00
900.00	5.00	0.02 (0.02)	0.49
1800.00	16.00	0.10 (0.10)	2.01
2700.00	39.00	0.28 (0.28)	5.64
3600.00	104.00	0.76 (0.75)	15.14
4500.00	322.00	2.17 (2.15)	56.71
5400.00	555.00	4.24 (4.23)	172.07
6300.00	722.00	6.50 (6.49)	279.93
7200.00	626.00	7.93 (7.92)	327.77
8100.00	481.00	8.55 (8.55)	346.58
9000.00	318.00	8.44 (8.44)	343.08
9900.00	162.00	7.69 (7.69)	320.67
10800.00	176.73	7.09 (7.09)	302.64
11700.00	191.45	6.63 (6.63)	285.10

ROUTING INTERVAL = 600.0 S

TIME (S)	INFLOW (M3/S)	RESERVOIR LEVEL (M)	OUTFLOW M3/S)
0.00	0.00	0.00	0.00
600.00	3.33	0.01	0.22
1200.00	8.67	0.04	0.78
1800.00	16.00	0.09	1.80
2400.00	31.33	0.19	3.76
3000.00	60.67	0.38	7.53
3600.00	104.00	0.69	13.87
4200.00	249.33	1.45	32.55
4800.00	399.67	2.57	77.87
5400.00	555.00	3.96	157.78
6000.00	666.33	5.42	231.13
6600.00	690.00	6.72	288.87
7200.00	626.00	7.67	320.22
7800.00	529.33	8.26	337.79
8400.00	426.67	8.51	345.23
9000.00	318.00	8.43	342.95
9600.00	214.00	8.07	332.14
10200.00	166.91	7.60	318.13
10800.00	176.73	7.20	306.01
11400.00	186.55	6.86	294.39

10-14

By Taylor's series

$$F(x_0+\Delta x, y_0+\Delta y) = F(x_0, y_0) + \frac{\partial F}{\partial x}\Delta x + \frac{\partial F}{\partial y}\Delta y + O(2) = 0 --- (1)$$

$$G(x_0+\Delta x, y_0+\Delta y) = G(x_0, y_0) + \frac{\partial G}{\partial x}\Delta x + \frac{\partial G}{\partial y}\Delta y + O(2) = 0 --- (2)$$

Solving eqs. 1 and 2 simultaneously for Δx and Δy

$$\Delta x = \frac{-F(x_0, y_0)\frac{\partial G}{\partial y} + G(x_0, y_0)\frac{\partial F}{\partial y}}{\frac{\partial F}{\partial x}\frac{\partial G}{\partial y} - \frac{\partial G}{\partial x}\frac{\partial F}{\partial y}}$$

$$\Delta y = \frac{-G(x_0, y_0)\frac{\partial F}{\partial x} + F(x_0, y_0)\frac{\partial G}{\partial x}}{\frac{\partial F}{\partial x}\frac{\partial G}{\partial y} - \frac{\partial G}{\partial x}\frac{\partial F}{\partial y}}$$

At critical flow

$$\frac{BQ^2}{gA^3} = 1$$

in which B = water surface width, Q = discharge, g = acceleration due to gravity and A = area of flow

```
C       COMPUTATION OF CRITICAL DEPTH IN A TRAPEZOIDAL CHANNEL BY
C         USING BISECTION METHOD
C
C       **********************NOTATION *************************
C
C
C       BO = CHANNEL-BOTTOM WIDTH;
C       BT = WATER-SURFACE WIDTH
C       G = ACCELERATION DUE TO GRAVITY;
C       Q = DISCHARGE;
C       S = SLOPE OF CHANNEL SIDES, S:HORIZONTAL TO 1 VERTICAL;
C       SO = CHANNEL BOTTOM SLOPE;
C       Y = CRITICAL DEPTH
C       YP, YN = DEPTH ESTIMATES SUCH THAT YP<Y<YN
C
        AR(Y)=(BO+S*Y)*Y
        BT(Y)=BO+2.*Y*S
        READ(5,*) G,Q,SO,S,BO,YP,YN
        WRITE(6,10)G,Q,SO,S,BO
 10     FORMAT(5X,'G =',F5.3,3X,'Q =',F8.3,' M3/S',3X,'SO =',F6.4,
       1 3X,'S =',F4.2,3X,'BO =',F6.2,' M')
        WRITE(6,15) YP,YN
 15     FORMAT(/5X,'INITIAL ESTIMATED FLOW DEPTHS:',2X, 'YP =',
       1   F6.2,' M',3X,'YN =',F4.2,' M')
        K = 0
 20     Y=0.5*(YP+YN)
        K=K+1
        IF (K.GT.50) GO TO 60
        F=1.0-(BT(Y)*Q*Q)/(G*AR(Y)**3.0)
        IF (F.LT.0.0) YN=Y
        IF (F.GT.0.0) YP=Y
        IF (ABS(YP-YN).LE.0.001) GO TO 40
        GO TO 20
 40     Y=0.5*(YP+YN)
        WRITE(6,50) Y,K
 50     FORMAT(/5X, 'CRITICAL DEPPTH =',F6.3,' M',5X,'NO. OF ITERATIONS
       *= ',1X,I2)
        GO TO 80
 60     WRITE(6,70)
 70     FORMAT(10X,'ITERATIONS FAILED')
 80     STOP
        END
```

```
G =9.810    Q = 110.000 M3/S    SO =0.0001    S =2.00    BO = 20.00 M

INITIAL ESTIMATED FLOW DEPTHS:   YP = 20.00 M    YN =0.20 M

CRITICAL DEPPTH = 1.387 M      NO. OF ITERATIONS   =  15
```

```
C
C     COMPUTATION OF CRITICAL DEPTH IN A TRAPEZOIDAL CHANNEL
C      USING NEWTON-RAPHSON METHOD
C
C     ******************** NOTATION *****************
C
C     BO = CHANNEL-BOTTOM WIDTH;
C     BT = WATER-SURFACE WIDTH;
C     G = ACCELERATION DUE TO GRAVITY;
C     Q = DISCHARGE;
C     S = SLOPE OF CHANNEL SIDES, S: HORIZONTAL TO 1 VERTICAL;
C     SO = CHANNEL-BOTTOM SLOPE;
C     Y = CRITICAL DEPTH;
C     YI = INITIAL ESTIMATE FOR CRITICAL DEPTH;
C
      AR(Y)= Y*(BO+S*Y)
      BT(Y) = BO+2.*Y*S
      READ(5,*) G,Q,BO,SO,S,YI
      WRITE(6,20) G,Q,BO,SO,S,YI
20    FORMAT(5X,'G =',F5.3,3X,'Q =',F7.3,' M3/S',3X,'BO =',
     1 F5.2,' M',3X,'SO =',F5.3,3X,'S =',F4.2,3X,'YI =',F5.2)
      K=0
30    K=K+1
      IF (K.GT.50) GO TO 80
      D=BT(YI)/AR(YI)*BT(YI)
      F=1.0-(BT(YI)*Q*Q)/(G*AR(YI)**3.0)
      FD=Q*Q/(G*AR(YI)**3.0)*(3.0*D-2.0*S)
      DY=F/FD
      IF (ABS(DY).LE.0.001) GO TO 60
      YI=YI-DY
      GO TO 30
60    WRITE(6,70) YI,K
70    FORMAT(5X,'CRITICAL DEPTH =',F6.2,' M',5X,'NO. OF ITERATIONS = ',
     *I2)
      GO TO 90
80    WRITE(6,85)
85    FORMAT(10X,'ITERATIONS FAILED')
90    STOP
      END
```

```
G =9.810   Q =110.000 M3/S   BO =20.00 M   SO =0.000   S =2.00   YI = 1.00
CRITICAL DEPTH =  1.39 M     NO. OF ITERATIONS =  5
```

Energy equation between sections I and I+1 can be written as

$$Y(I) + Z(I) + \frac{Q^2}{2gA(I)^2} = Y(I+1)^2 + \frac{Q^2}{2gA(I+1)^2} + Z(I+1)^2 + \frac{\Delta x}{2}[SF(I) + SF(I+1)] \text{-------} (1)$$

in which, SF is slope of the energy grade line.

$$SF = \frac{Q^2 n^2 P^{1.333}}{A^{3.333}} \text{-----------------------------------} (2)$$

in which A = cross sectional area and P = wetted perimeter. E quation 1 can be written for sections 1 to N giving N equations. In these N equations, since A(I) and P(I) are functions of Y(I), the only unknowns are Y(I), I = 1, N+1. Therefore, there are (N+1) unknowns in N equations. The extra equation needed needed to solve is given by

$$Y(N+1) = YDS \qquad (3)$$

in which, YDS is the downstream depth known from the boundary condition for subcritical flow.

Above system of equation is non-linear and hence, may be solved by Newton-Raphson technique for simultaneous equations as discussed in the text.

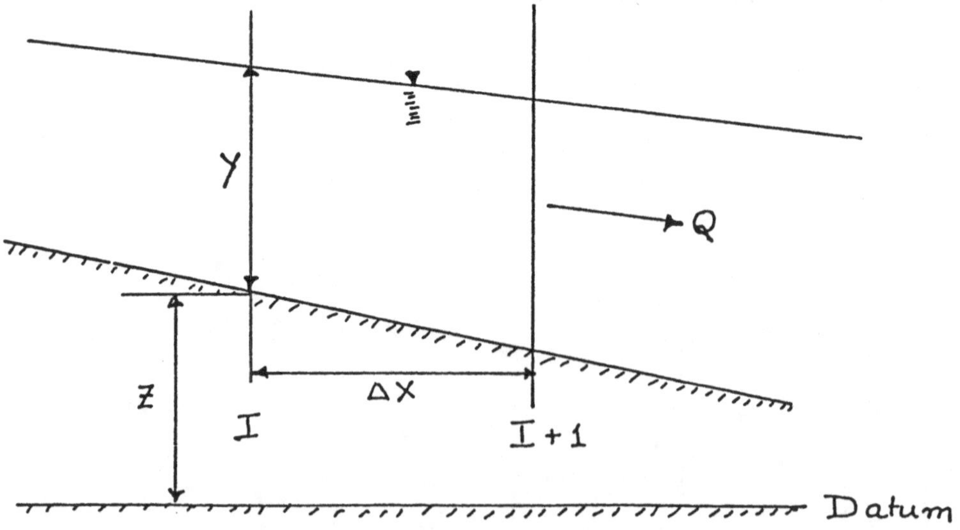

```
C
C----------------------------------------------------------------------
C
C        SIMULTANEOUS SOLUTION FOR BACK-WATER PROFILE
C
C--------------------NOTATION------------------------------------------
C
C        A = AREA OF CROSS SECTION;
C        B = WATER-SURFACE WIDTH;
C        BB = BOTTOM WIDTH;
C        CL = CHANNEL LENGTH;
C        C1,C2 = COEFFICIENT MATRIX ELEMENTS FORSOLVING DY(I);
C        F = FUNCTIONAL EQUATIONS THAT ARE SOLVED;
C        G = ACCELERATION DUE TO GRAVITY;
C        I = NODE NUMBER
C        N = NUMBER OF REACHES;
C        P = WETTED PERIMETER;
C        Q = DISCHARGE;
C        ROUGN = MANNINGS COEFFICIENT;
C        SF = SLOPE OF ENERGY GRADE LINE;
C        SS = SIDE SLOPE;
C        SO = BOTTOM SLOPE;
C        YDS = DEPTH AT DOWNSTREAM POINT;
C
C----------------------------------------------------------------------
      DIMENSION Y(50),A(50),B(50),P(50),SF(50),F(50),C1(50),C2(50),
     *DY(50)
      READ(5,*)Q,ROUGN,BB,SS,SO,YDS,CL,N,G
      WRITE(6,1)Q,ROUGN,BB,SS,SO,YDS,CL,N,G
1     FORMAT(5X,'Q = ',F8.2,1X,'M3/S',3X,'MAN. N = ',F8.3,3X,
     *'BOT. WIDTH = ',F8.2,1X,'M',3X//
     *5X,'SIDE SLOPE = ',F3.1,3X,'BOT. SLOPE = ',F8.4,3X,
     *'YDS = ',F6.2,1X,'M'//5X,'LENGTH = ',F8.2,1X,'M',3X,'N = ',I2,
     *3X,'G = ',F8.2,1X,'M/S2')
      N1=N+1
      READ(5,*)(Y(I),I=1,N1)
      WRITE(6,2)(Y(I),I=1,N1)
2     FORMAT(/5X,'INITIAL ESTIMATED DEPTHS'//5X,8F5.1/5X,8F5.1)
C
C
      DX=CL/N
      IFLAG=0
5     CONTINUE
      IF(IFLAG.GE.50) GO TO 998
      DO 10 I=1,N1
      A(I)=(BB+SS*Y(I))*Y(I)
      B(I)=BB+SS*2.0*Y(I)
      P(I)=BB+2.0*Y(I)*SQRT(SS*SS+1.0)
      SF(I)=((Q*ROUGN)**2.0)*(P(I)**1.333)/(A(I)**3.333)
10    CONTINUE
      DO 30 I=1,N
C
C
      F(I)=Y(I)-Y(I+1)+SO*DX+(Q*Q/(2.0*G))*((1.0/(A(I)**2.0))-
     *(1.0/(A(I+1)**2.0)))


      F(I)=F(I)-0.5*DX*(SF(I)+SF(I+1))
      F(I)=-F(I)
30    CONTINUE
C
C
C        CALCULATION OF ELEMENTS OF COEFFICIENT MATRIX
C
C
      DO 40 I=1,N
      C1(I)=1.0-Q*Q*B(I)/(G*A(I)**3.0)-(DX*SF(I)/3.0)*
     *((4.0*SQRT(SS*SS+1.0)/P(I))-5.0*B(I)/A(I))
      C2(I)=-1.0+Q*Q*B(I+1)/(G*A(I+1)**3.0)-(DX*SF(I+1)/3.0)*
     *((4.0*SQRT(SS*SS+1.0)/P(I+1))-5.0*B(I+1)/A(I+1))
40    CONTINUE
C
C
C        CALCULATION OF DY(I)
C
```

```
C
      DY(N1)=0.0
      DO 50 I=N,1,-1
      DY(I)=(F(I)-C2(I)*DY(I+1))/C1(I)
50    CONTINUE
      SUM=0.0
      DO 60 I=1,N1
      SUM=SUM+ABS(DY(I))
60    CONTINUE
      DO 70 I=1,N1
      Y(I)=Y(I)+DY(I)
70    CONTINUE
      IF(SUM.LE.0.01) GO TO 100
      IFLAG=IFLAG+1
      GO TO 5
100   CONTINUE
      DO 1300 I=1,N1
      X=(I-1)*DX
      WRITE(6,1200) X,Y(I)
1200  FORMAT(/5X,'X = ',F9.4,1X,'M',3X,'Y = ',F8.4,1X,'M')
1300  CONTINUE
      GO TO 999
998   WRITE(6,*) 'ITER TROUBLE'
999   STOP
      END
```

RESULT FILE FOR EXAMPLE 10.4

```
Q =    110.00 M3/S   MAN. N =    0.013   BOT. WIDTH =    20.00 M

SIDE SLOPE = 2.0   BOT. SLOPE =   0.0001   YDS =   5.00 M

LENGTH =   1500.00 M   N = 15   G =     9.81 M/S2



 INITIAL ESTIMATED DEPTHS

   5.0  5.0  5.0  5.0  5.0  5.0  5.0  5.0
   5.0  5.0  5.0  5.0  5.0  5.0  5.0  5.0

 X =     0.0000 M   Y =   4.8746 M

 X =   100.0000 M   Y =   4.8829 M

 X =   200.0000 M   Y =   4.8912 M

 X =   300.0000 M   Y =   4.8995 M

 X =   400.0000 M   Y =   4.9078 M

 X =   500.0000 M   Y =   4.9161 M

 X =   600.0000 M   Y =   4.9245 M

 X =   700.0000 M   Y =   4.9328 M

 X =   800.0000 M   Y =   4.9412 M

 X =   900.0000 M   Y =   4.9496 M

 X =  1000.0000 M   Y =   4.9579 M

 X =  1100.0000 M   Y =   4.9663 M

 X =  1200.0000 M   Y =   4.9747 M

 X =  1300.0000 M   Y =   4.9831 M

 X =  1400.0000 M   Y =   4.9916 M

 X =  1500.0000 M   Y =   5.0000 M
```

The error first decreased because of reduction in truncation errors. The error increases for a higher number of sub-intervals because the round-off errors are large in that case.

```
C--------------------------------------------------------------------------
C         TRAPEZOIDAL RULE ERROR CALCULATION
C
C--------------------------------------------------------------------------
      WRITE(6,30)
      WRITE(6,20)
      WRITE(6,30)
      PI =4.0*ATAN(1.0)
20    FORMAT(4X,'TRAPEZOIDAL RULE ERROR CALCULATION')
30    FORMAT(4X,'-----------------------------------')
      WRITE(6,40)
40    FORMAT(4X,'NO OF INTER      SUM      ERROR')
      WRITE(6,30)
      N=1
1     X=PI/N
      SUM = 0.0
      DO 3 I=1,N+1
      P=(I-1)*X
      Y=SIN(P)
      IF(I.EQ.1.OR.I.EQ.(N+1)) GO TO 2
      SUM=SUM+2.0*Y
      GO TO 3
2     SUM=SUM+Y
3     CONTINUE
      SUMDI=(X/2.0)*SUM
      ERR=((2.0-SUMDI)/2.0)*100.0
      WRITE(6,10) N,SUMDI,ERR
10    FORMAT(5X,I5,6X,2F8.4)
      N=2.0*N
      IF(N.LT.2500) GO TO 1
      WRITE(6,30)
      STOP
      END
```

TRAPEZOIDAL RULE ERROR CALCULATION

NO OF INTER	SUM	ERROR
1	0.0000	99.9999
2	1.5708	21.4602
4	1.8961	5.1941
8	1.9742	1.2885
16	1.9936	0.3217
32	1.9984	0.0806
64	1.9996	0.0205
128	1.9999	0.0055
256	1.9999	0.0049
512	1.9999	0.0061
1024	1.9999	0.0072
2048	1.9998	0.0079

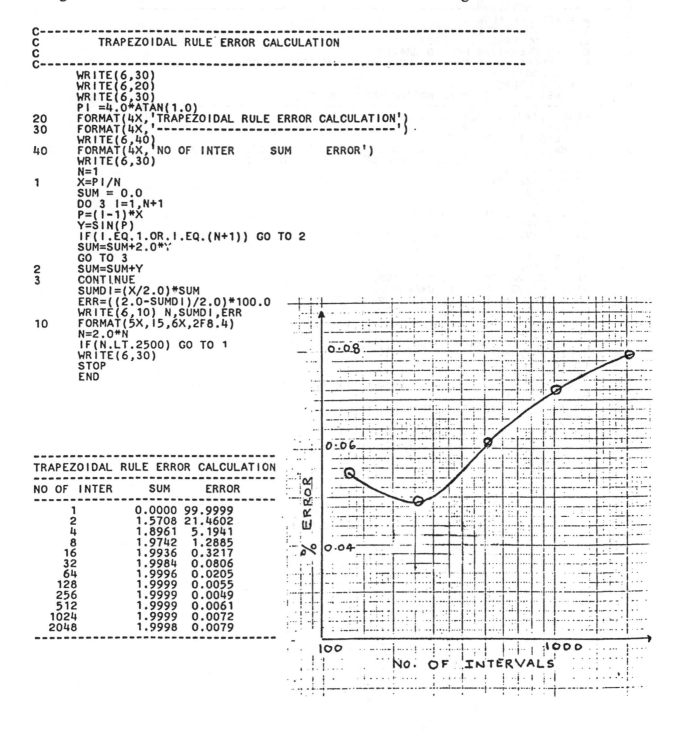

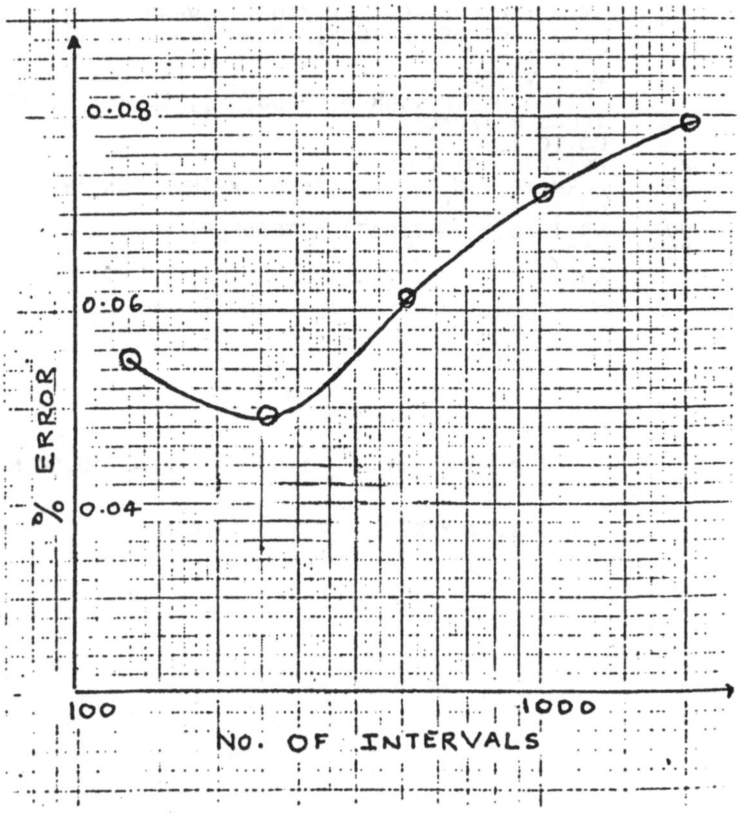

```
C-----------------------------------------------------------------
C             CHANNEL TRANSITION
C
C------------NOTATION---------------------------------------------
C     A = AREA OF FLOW ;
C     B = WIDTH OF CROSS-SECTION ;
C     ERR = TOLERANCE IN ERROR ;
C     G = ACCELERATION DUE TO GRAVITY ;
C     RK = COEFFICIENT OF HEAD LOSSES AT THE TRANSITION ;
C     V = VELOCITY ;
C     Y = FLOW DEPTH ;
C     Z = HEIGHT OF CHANNEL BOTTOM ABOVE DATUM ;
C-----------------------------------------------------------------
      READ(5,*)Y1,V1,B,Z1,RK,G,ERR
      READ(5,*) Z2,Y2,V2
      WRITE(6,5) Y1,V1,B,Z2
5     FORMAT(5X,'Y1 = ',F6.2,1X,'M',2X,'V1 = ',F6.2,1X,'M/S',2X,
     *'B = ',F6.2,1X,'M',2X,'STEP RISE = ',F6.2,1X,'M')
      C1 = B*Y1*V1
      C2 = Z1-Z2+Y1+V1*V1/(2.0*G)
      IFLAG = 0
10    CONTINUE
      IF(IFLAG.GE.50) GO TO 998
      FY = B*Y2*V2-C1
      GY = Y2+V2*V2/(2.0*G)+RK*ABS(V1*V1-V2*V2)/(2.0*G)-C2
      DFY = B*V2
      DGY = 1.0
      DFV = B*Y2
      IF(V1.GE.V2) DGV = (1.0-RK)*V2/G
      IF(V1.LT.V2) DGV = (1.0+RK)*V2/G
      DY = (-FY*DGV+GY*DFV)/(DFY*DGV-DGY*DFV)
      DV = (-GY*DFY+FY*DGY)/(DFY*DGV-DGY*DFV)
      Y2 = Y2+DY
      V2 = V2+DV
      IF(ABS(DY).LE.ERR.AND.ABS(DV).LE.ERR) GO TO 20
      IFLAG = IFLAG+1
      GO TO 10
20    WRITE(6,30) V2,Y2
30    FORMAT(//5X,'V2 = ',F6.2,1X,'M/S',5X,'Y2 = ',F6.2,1X,'M')
      GO TO 999
998   WRITE(8,*) 'ITERATION TROUBLE'
999   STOP
      END
```

```
     Y1 =   2.00 M  V1 =   3.00 M/S  B =  10.00 M  STEP RISE =   0.10 M

     V2 =   3.37 M/S     Y2 =   1.78 M
```

10-20

PROBLEM 4.6 BISECTION METHOD

N = 0.013 Q = 3000.000 M3/S SO = 0.0020 S = 2.00 BO = 10.00 M
INITIAL ESTIMATED FLOW DEPTHS: YP = 0.20 M YN = 15.00 M
NORMAL DEPTH = 11.319 M NO. OF ITERATIONS = 14

PROBLEM 4.6 NEWTON-RAPHSON METHOD

N = 0.013 Q = 3000.0 M3/S BO = 10.00 M SO = 0.001 S = 2.00 YI = 5.00
NORMAL DEPTH = 11.32 M NO. OF ITERATIONS = 5

PROBLEM 4.9 BISECTION METHOD

N = 0.013 Q = 1000.000 M3/S SO = 0.0020 S =1.00 BO = 10.00 M
INITIAL ESTIMATED FLOW DEPTHS: YP = 0.20 M YN = 15.00 M
NORMAL DEPTH = 6.890 M NO. OF ITERATIONS = 14

PROBLEM 4.9 NEWTON-RAPHSON METHOD

N = 0.013 Q = 1000.0 M3/S BO = 10.00 M SO = 0.002 S = 1.00 YI = 5.0
NORMAL DEPTH = 6.89 M NO. OF ITERATIONS = 4

10-21

$$Y_1 + \frac{Q^2}{2gA_1^{\ 2}} = Y_2 + \frac{Q^2}{2gA_2^{\ 2}}$$

in which A = flow area, Q = discharge, and Y = depth

```
C--------------------------------------------------------------------
C         COMPUTATION OF ALTERNATE DEPTH
C
C
C--------------------------------------------------------------------
      READ(5,*) Y1,A1,Q,G,ERR,BO,S
      READ(5,*) Y2
      WRITE(6,50) Y1,BO,S,Q
50    FORMAT(5X,'Y1 = ',F6.2,1X,'M',5X,'BO = ',F6.2,1X,'M',5X,
     *  'S = ',F6.2,5X,'Q = ',F6.3,1X,'M3/S')
      C1 = Y1+Q*Q/(2.0*G*A1*A1)
      C2 = Q*Q/(2.0*G)
      IFLAG = 0
10    CONTINUE
      IF(IFLAG.GT.50) GO TO 998
      B2 = BO+2.0*S*Y2
      A2 = Y2*(BO+S*Y2)
      FY = Y2+C2/(A2*A2)-C1
      DFY = 1.0-2.0*C2*B2/(A2**3.0)
      DY2 = -FY/DFY
      Y2 = Y2+DY2
      IF(ABS(DY2).LE.ERR) GO TO 20
      IFLAG = IFLAG+1
      GO TO 10
20    WRITE(6,30)Y2
30    FORMAT(//5X,'ALTERNATE DEPTH = ',F6.2,1X,'M')
      GO TO 999
998   WRITE(6,*) 'ITERATION TROUBLE'
999   STOP
      END
```

Y1 = 0.10 M BO = 1.00 M S = 0.00 Q = 0.600 M3/S

ALTERNATE DEPTH = 1.93 M

10-22

FOLLOWING CHANGES ARE MADE IN THE PROGRAMS FOR COMPUTING AREA AND TOP WIDTH

```
AR(Y) = 3.142*RA*RA(ATAN(SQRT(RA*RA-(RA-Y)**2.0)/(RA-Y))/3.142)
        -SQRT(RA*RA-(RA-Y)**2.0)*RA-Y)
BT(Y) = 2.0*SQRT(RA*RA-(RA-Y)**2.0)
```

BISECTION METHOD

G = 32.200 Q = 25.000 M3/S RA = 2.00
INITIAL ESTIMATED FLOW DEPTHS: YP = 2.00 M YN = 0.20 M
CRITICAL DEPTH = 1.477 M NO. OF ITERATIONS = 11

NEWTON-RAPHSON METHOD

G = 32.200 Q = 25.000 M3/S RA = 2.00 YI = 1.00
CRITICAL DEPTH = 1.48 M NO. OF ITERATIONS = 6

10-23

PROBLEM 4.30 STEP RISE OF 0.3 M
Y1 = 3.00 M V1 = 3.00 M/S B = 1.00 M STEP RISE = 0.30 M
V2 = 3.61 M/S Y2 = 2.50 M

CHANGE IN DEPTH = 2.50 – 3.00 = - 0.5 M
CHANGE IN WATER-SURFACE ELEVATION = 2.5 + 0.3 – 3.0 = -0.2 M

PROBLEM 4.30 STEP DOWN OF 0.3 M
Y1 = 3.00 M V1 = 3.00 M/S B = 1.00 M STEP DOWN = 0.30 M

V2 = 2.65 M/S Y2 = 3.40 M

CHANGE IN DEPTH = 3.4 – 3.0 = 0.40 M
CHANGE IN WATER-SURFACE ELEVATION = 3.4 – 0.3 – 3.0 = 0.1 M

PROBLEM 4.31 STEP RISE OF 0.6 M
Y1 = 3.00 M V1 = 2.00 M/S B = 1.00 M STEP RISE = 0.60 M
V2 = 2.68 M/S Y2 = 2.24 M

CHANGE IN DEPTH = 2.24 – 3.00 = - 0.76 M

PROBLEM 4.31 STEP DOWN OF 0.15 M
Y1 = 3.00 M V1 = 2.00 M/S B = 1.00 M STEP DOWN = 0.15 M
V2 = 1.89 M/S Y2 = 3.17 M

CHANGE IN DEPTH = 3.17 – 3.00 = 0.17 M

10-24

```
C
C       RESERVOIR ROUTING USING IMPROVED EULER METHOD
C                  WITH PARABOLIC INTERPOLATION
C
C       ************** NOTATION ****************
C
C       AR(I) = SURFACE AREA OF RESERVOIR AT LEVEL ELAR(I);
C       DT = ROUTING INTERVAL;
C       NAR = NUMBER OF POINTS ON THE STAGE VS. AREA CURVE;
C       NQO = NUMBER OF POINTS ON THE STAGE VS. OUTFLOW CURVE;
C       NT = NUMBER OF POINTS ON THE INFLOW VS TIME CURVE;
C       QIN(I) = INFLOW AT TIME(I);
C       QO(I) = OUTFLOW AT WATER LEVEL ELQO(I);
C       QO1 = OUTFLOW AT TIME T = 0;
C       Z = RESERVOIR LEVEL ABOVE DATUM;
C       Z1 = RESERVOIR LEVEL AT TIME T = 0;
C       TSTOP = TIME UPTO WHICH ROUTING IS TO BE COMPUTED.
C
        DIMENSION AR(50),ELAR(50),ELQO(50),QO(50),TIME(50),QIN(50)
C
C       INITIAL CONDITIONS
C
        READ(5,*) Z1,QO1,DT,TSTOP
        WRITE(6,20) Z1,QO1,DT,TSTOP
20      FORMAT(//5X,'INITIAL RESERVOIR WATER LEVEL =',F7.2,' M'/
       1 5X,'INITIAL OUTFLOW =',F5.1,' M3/S'/
       2 5X,'ROUTING INTERVAL =',F6.1,' S'/
       3 5X,'TIME UPTO WHICH ROUTING IS TO BE DONE =',F8.1,' S'/)
C
C       STAGE VS. RESERVOIR-SURFACE AREA CURVE
C
        READ (5,*) NAR,(ELAR(I),AR(I),I=1,NAR)
        WRITE(6,30)
30      FORMAT(15X,'STAGE',5X,'RESERVOIR SURFACE AREA'/
       1  16X,'(M)',13X,'(SQ. M)')
        WRITE(6,40) (ELAR(I),AR(I),I=1,NAR)
40      FORMAT(9X,F10.1,10X,F10.1)
C
C       STAGE-OUTFLOW CURVE
C
        READ (5,*) NQO,DQO,(ELQO(I),QO(I),I=1,NQO)
        WRITE(6,60)
60      FORMAT(/15X,'STAGE',9X,'  OUTFLOW'/
       1  16X,'(M)',12X,'(M3/S)')
        WRITE(6,70) (ELQO(I),QO(I),I=1,NQO)
70      FORMAT(9X,F10.1,8X,F10.1)
C
C       INFLOW HYDROGRAPH
C
        READ(5,*) NT, (TIME(I),QIN(I), I=1,NT)
        WRITE(6,90)
90      FORMAT(/14X,'TIME',12X,'INFLOW'/
       1  14X,'(S)',13X,'(M3/S)')
        WRITE(6,100) (TIME(I),QIN(I),I=1,NT)
100     FORMAT(8X,F10.1,8X,F10.2)


        T = 0.
        WRITE(6,125)
125     FORMAT(//13X,'TIME',11X,'INFLOW',4X,'RESERVOIR LEVEL',2X,
       1   'OUTFLOW'/13X,'(S)',12X,'(M3/S)',9X,'(M)',11X,'M3/S)')
        WRITE(6,240) T,QIN(1),Z1,QO1
C
C
135     CONTINUE
        DO 140 I=1,NT
        IF (T.LT.TIME(I)) GO TO 150
140     CONTINUE
150     I1=I-1
        QIP=QIN(I1)+(T-TIME(I1))/(TIME(I)-TIME(I1))*(QIN(I)-QIN(I1))
        DO 160 I=1,NQO
        IF (Z1.LT.ELQO(I)) GO TO 170
160     CONTINUE
170     I1=I-1
        CALL PARAB(I1,Z1,ELQO,QO,QOP,DQO)
        DO 180 I=1,NAR
        IF (Z1.LT.ELAR(I)) GO TO 190
180     CONTINUE
190     I1=I-1
        ARP=AR(I1)+(Z1-ELAR(I1))/(ELAR(I)-ELAR(I1))*(AR(I)-AR(I1))
        DZP1=(QIP-QOP)/ARP
        ZP1=Z1+DZP1*DT
```

```
C
      T=T+DT
      DO 640 I=1,NT
      IF (T.LT.TIME(I)) GO TO 650
640   CONTINUE
650   I1=I-1
      QIP=QIN(I1)+(T-TIME(I1))/(TIME(I)-TIME(I1))*(QIN(I)-QIN(I1))
      DO 660 I=1,NQO
      IF (ZP1.LT.ELQO(I)) GO TO 670
660   CONTINUE
670   I1=I-1
      CALL PARAB(I1,ZP1,ELQO,QO,QOP,DQO)
      DO 680 I=1,NAR
      IF (ZP1.LT.ELAR(I)) GO TO 690
680   CONTINUE
690   I1=I-1
      ARP=AR(I1)+(ZP1-ELAR(I1))/(ELAR(I)-ELAR(I1))*(AR(I)-AR(I1))
      DZP2=(QIP-QOP)/ARP
      Z2=Z1+0.5*(DZP1+DZP2)*DT
C
C
      DO 232 I=1,NQO
      IF (Z2.LT.ELQO(I)) GO TO 235
232   CONTINUE
235   I1=I-1
      CALL PARAB(I1,Z2,ELQO,QO,Q2,DQO)
      WRITE(6,240) T,QIP,Z2,Q2
240   FORMAT(9X,F10.2,5X,F10.2,2X,F10.2,5X,F10.2)
      IF (T.GT.TSTOP) GO TO 250
      Z1=Z2



      GO TO 135
250   STOP
      END
      SUBROUTINE PARAB(I1,ZZ,ELQO,QO,QOZ,DQO)
      DIMENSION ELQO(50),QO(50)
      R = (ZZ-ELQO(I1))/DQO
      IF(I1.EQ.1) R=R-1.0
      IF(I1.LT.2) I1=2
      QOZ = QO(I1)+0.5*R*(QO(I1+1)-QO(I1-1)+R*(QO(I1+1)+QO(I1-1)
     *-2.0*QO(I1)))
      RETURN
      END
```

```
INITIAL RESERVOIR WATER LEVEL =    0.00 M
INITIAL OUTFLOW =  0.0 M3/S
ROUTING INTERVAL = 900.0 S
TIME UPTO WHICH ROUTING IS TO BE DONE =  9900.0 S
```

STAGE (M)	RESERVOIR SURFACE AREA (SQ. M)
0.0	176400.0
1.0	179800.0
2.0	183200.0
3.0	186600.0
4.0	190100.0
5.0	193600.0
6.0	197100.0
7.0	200700.0
8.0	204300.0
9.0	207900.0
10.0	211600.0

STAGE (M)	OUTFLOW (M3/S)
0.0	0.0
1.0	20.0
2.0	48.0
3.0	100.0
4.0	160.0
5.0	210.0
6.0	260.0
7.0	300.0
8.0	330.0
9.0	360.0
10.0	380.0

TIME (S)	INFLOW (M3/S)
0.0	0.00
900.0	5.00
1800.0	16.00
2700.0	39.00
3600.0	104.00
4500.0	322.00
5400.0	555.00
6300.0	722.00
7200.0	626.00
8100.0	481.00
9000.0	318.00
9900.0	162.00
10800.0	100.00

TIME (S)	INFLOW (M3/S)	RESERVOIR LEVEL (M)	OUTFLOW M3/S)
0.00	0.00	0.00	0.00
900.00	5.00	0.01	0.20
1800.00	16.00	0.06	1.04
2700.00	39.00	0.20	3.28
3600.00	104.00	0.53	9.67
4500.00	322.00	1.53	33.74
5400.00	555.00	3.35	119.84
6300.00	722.00	5.52	235.98
7200.00	626.00	7.31	310.29
8100.00	481.00	8.29	338.80
9000.00	318.00	8.52	345.58
9900.00	162.00	8.07	331.96
10800.00	100.00	7.22	307.52

10-25

RESERVOIR ROUTING USING EULER METHOD

RESULT FILE (INPUT: SAME AS FOR PROBLEM 10.2)

ROUTING INTERVAL(S)	PEAK OF OUTFLOW (M3/S)
600	356.9
700	360.0
800	361.0
900	364.7
1000	363.0
1100	364.5
1200	367.0

PEAK OUTFLOW IN GENERAL INCREASES WITH AN INCREASE IN ROUTING INTERVAL.

10-26

```
C
C       COMPUTATION OF WATER-SURFACE PROFILE BY USING
C                    EULER METHOD
C
C       *************** NOTATION *****************
C       A=FLOW AREA;
C       B=TOP WATER-SURFACE WIDTH;
C       BO=CHANNEL-BOTTOM WIDTH;
C       P = WETTED PERIMETER;
C       Q = DISCHARGE;
C       MN = MANNING'S N;
C       S = CHANNEL-SIDE SLOPE, S HORIZONTAL : 1 VERTICAL;
C       SO = CHANNEL-BOTTOM SLOPE;
C       X = DISTANCE ALONG CHANNEL BOTTOM, POSITIVE IN THE DOWNSTREAM
C            DIRECTION;
C       Y = FLOW DEPTH
C       YD = DEPTH AT DOWNSTREAM END.
C
        REAL MN
        DIMENSION X(100)
        AR(Y)=Y*(BO+S*Y)
        WP(Y)=BO+2.*Y*SQRT(1.+S*S)
        READ (5,*) BO,S,SO,MN,Q,YD
        READ (5,*) N,(X(I),I=1,N)
        WRITE(6,10) BO,S,SO,MN,Q,YD
10      FORMAT(2X,'B =',F5.1,' M',2X,'S =',F4.1,2X,'SO =',F6.4,
       1   2X,'N =',F5.3,2X,'Q =',F9.1,' M3/S',2X,'YD =',F7.3,' M')
        Q2=Q*Q
        QN2=(MN*Q)**2
        Y=YD
        WRITE(6,15)
15      FORMAT(6X,'X',10X,'Y')
        WRITE(6,20) X(1),Y
        DO 30 I = 2,N
        DX=X(I)-X(I-1)
        A=AR(Y)
        P=WP(Y)
        R=A/P
        SF1=QN2/(A*A*R**1.333)
        B=BO+2.*S*Y
        DY1=(SO-SF1)/(1-(B*Q2)/(32.2*A**3))
        Y=Y+DY1*DX
        WRITE(6,20) X(I),Y
20      FORMAT(F10.1,F10.3)
30      CONTINUE
        STOP
        END
```

INPUT FILE

```
200.0,0.0,0.004,0.013,20000.0,50.0
16,0.0,-50.0,-100.0,-200.0,-300.0,-500.0,-800.0,-1200.0,-1600.0,
-2000.0,-2500.0,-3000.0,-3500.0,-4000.0,-4500.0,-5000.0
```

OUTPUT FILE

```
 B =200.0 ft  S = 0.0   SO =0.0040   N =0.013   Q =  20000.0 cfs   YD = 50.ft
      X          Y
     0.0      50.000
   -50.0      49.800
  -100.0      49.600
  -200.0      49.199
  -300.0      48.799
  -500.0      47.998
  -800.0      46.797
 -1200.0      45.195
 -1600.0      43.593
 -2000.0      41.991
 -2500.0      39.988
 -3000.0      37.984
 -3500.0      35.980
 -4000.0      33.975
 -4500.0      31.968
 -5000.0      29.961
```

```
C
C      COMPUTATION OF WATER-SURFACE PROFILE BY USING
C                    RUNGE-KUTTA METHOD
C
C      *************** NOTATION *****************
C      A=FLOW AREA;
C      B=TOP WATER-SURFACE WIDTH;
C      BO=CHANNEL-BOTTOM WIDTH;
C      P = WETTED PERIMETER;
C      Q = DISCHARGE;
C      MN = MANNING'S N;
C      S = CHANNEL-SIDE SLOPE, S HORIZONTAL : 1 VERTICAL;
C      SO = CHANNEL-BOTTOM SLOPE;
C      X = DISTANCE ALONG CHANNEL BOTTOM, POSITIVE IN THE DOWNSTREAM
C            DIRECTION;
C      Y = FLOW DEPTH
C      YD = DEPTH AT DOWNSTREAM END.
C
       REAL MN
       DIMENSION X(100)
       AR(Y)=Y*(BO+S*Y)
       WP(Y)=BO+2.*Y*SQRT(1.+S*S)
       READ (5,*) BO,S,SO,MN,Q,YD
       READ (5,*) N,(X(I),I=1,N)
       WRITE(6,10) BO,S,SO,MN,Q,YD
10     FORMAT(2X,'B =',F5.1,' M',2X,'S =',F4.1,2X,'SO =',F6.4,
     1   2X,'N =',F5.3,2X,'Q =',F9.1,' M3/S',2X,'YD =',F7.3,' M')
       Q2=Q*Q
       QN2=(MN*Q)**2
       Y=YD
       WRITE(6,15)
15     FORMAT(6X,'X',10X,'Y')
       WRITE(6,20) X(1),Y
       DO 30 I = 2,N
       DX=X(I)-X(I-1)
       A=AR(Y)
       P=WP(Y)
       R=A/P
       SF1=QN2/(A*A*R**1.333)
       B=BO+2.*S*Y
       DY1=(SO-SF1)/(1-(B*Q2)/(32.2*A**3))
       Y2=Y+DY1*DX*0.5
       A=AR(Y2)
       P=WP(Y2)
       R=A/P
       SF1=QN2/(A*A*R*1.333)
       B=BO+2.0*S*Y
       DY2=(SO-SF1)/(1-(B*Q2)/(32.2*A**3))
       Y2=Y+DY2*DX*0.5
       A=AR(Y2)
       P=WP(Y2)
       R=A/P
       SF1=QN2/(A*A*R*1.333)
       B=BO+2.0*S*Y
       DY3=(SO-SF1)/(1-(B*Q2)/(32.2*A**3))
       Y2=Y+DY3*DX
       A=AR(Y2)
       P=WP(Y2)
       R=A/P
       SF1=QN2/(A*A*R*1.333)
       B=BO+2.0*S*Y
       DY4=(SO-SF1)/(1-(B*Q2)/(32.2*A**3))
       Y=Y+(DX/6.0)*(DY1+2.0*(DY2+DY3)+DY4)
       WRITE(6,20) X(I),Y
20     FORMAT(F10.1,F10.3)
30     CONTINUE
       STOP
       END
```

OUTPUT FILE

```
B =200.0 M  S = 0.0  SO =0.0040  N =0.013  Q =  20000.0 M3/S  YD = 50
      X          Y
     0.0      50.000
   -50.0      49.800
  -100.0      49.600
  -200.0      49.201
  -300.0      48.801
  -500.0      48.002
  -800.0      46.803
 -1200.0      45.205
 -1600.0      43.607
 -2000.0      42.009
 -2500.0      40.012
 -3000.0      38.015
 -3500.0      36.018
```

11-1

D = 2 m, e = 20 mm, E = 206 Gpa, K = 2.19 Gpa, ρ = 999 kg/m3
Using Eq. 11-18

$$a = \sqrt{\left(\frac{2.19*10^9}{999}\right) / \left(1 + \frac{2.19*2}{20*10^{-3}*206}\right)} = 1031 m/s$$

Initial velocity in the pipe = $\frac{10*4}{\pi(2)^2}$ = 3.183 m/s
Δv = 0.0 - 3.183 = -3.183 m/s
ΔH = -(a/g)Δv = -(1031/9.81)* (-3.183) = 334.5 m
Pressure rise caused by instantaneous closure = 334.5 m

11-2

From problem 11-1, l = 1962 m, a = 1031 m/s, ΔH = 334.5 m, Ho = 100 m.

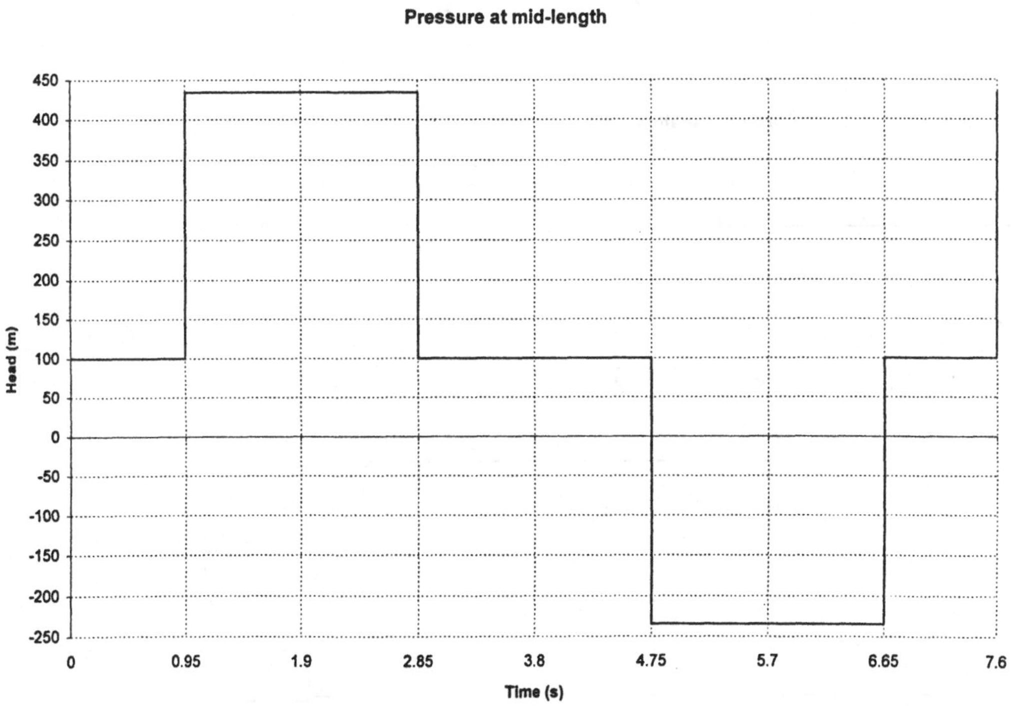

11-3

a = 1000 m/s, ΔH = (a/g)*Δv = (1000/9.81)*1.0 = 102 m

Upstream pipe length = L_u = 2000 m

Downstream pipe length = L_d = 4000 m

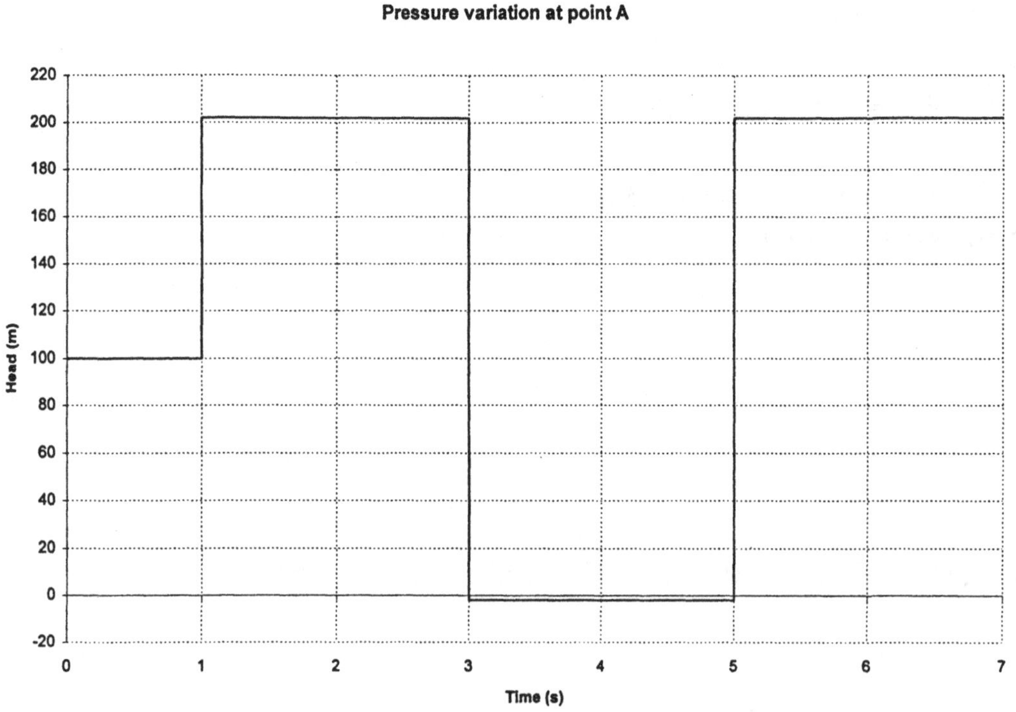

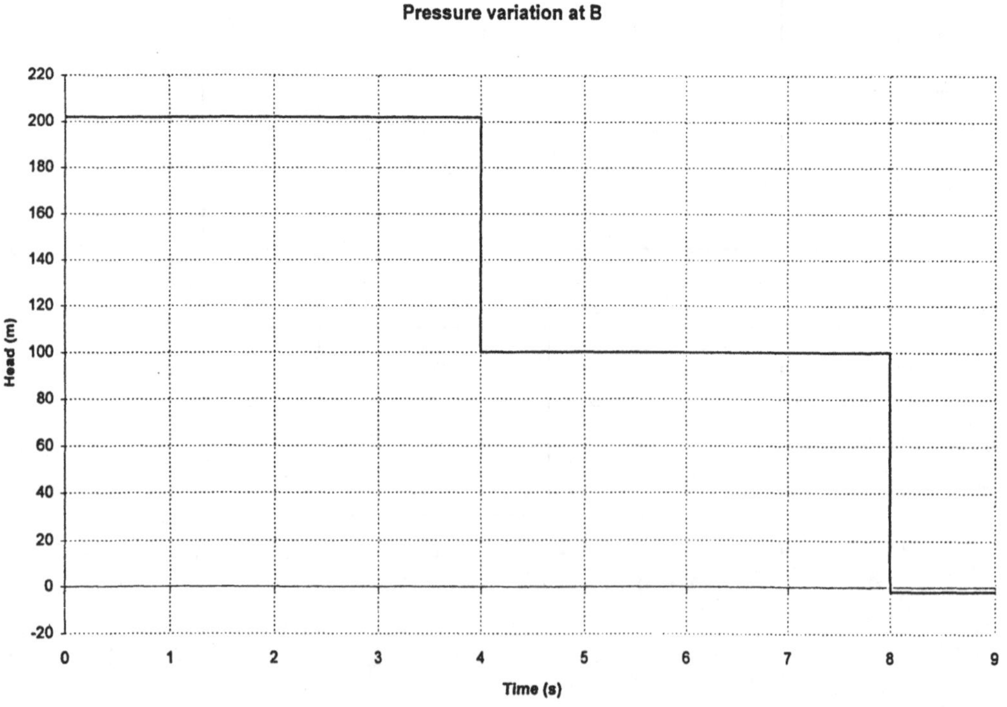

Pressure variation at C

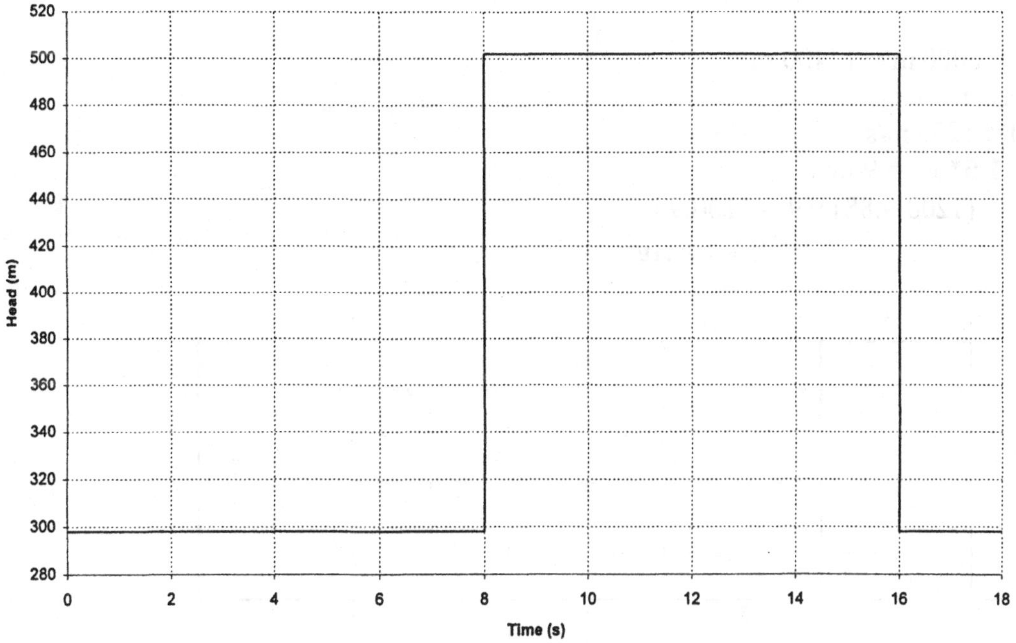

Pressure variation at D

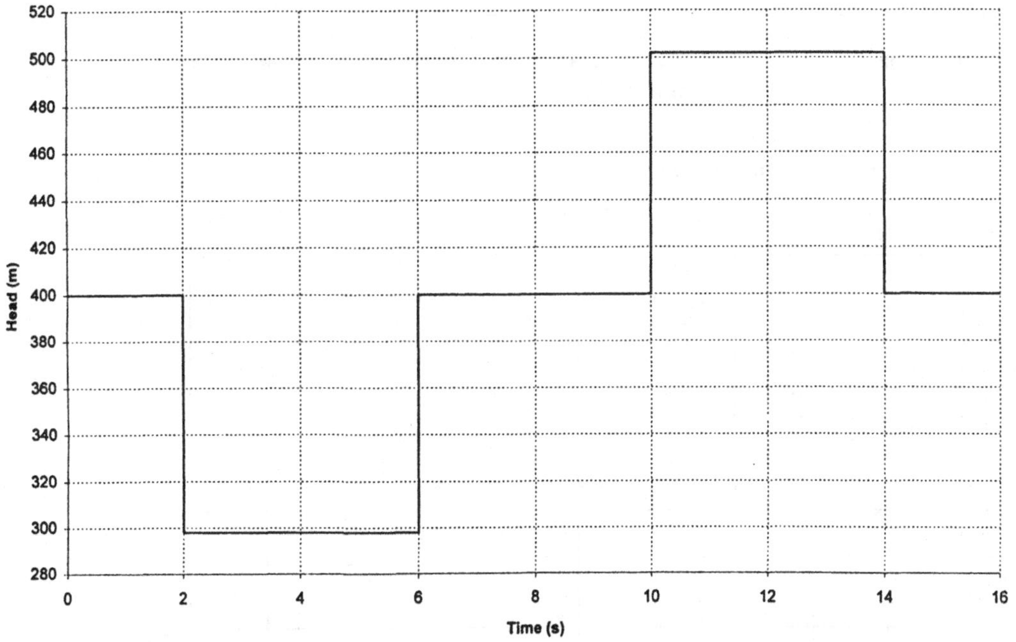

11-4

Take M = 6

L = 2.4*M = 2.4*6000 m = 14400 m

a = 1.2*r where r = 1

a = 1.2*1*1000 = 1200 m/s

Vo = 1.5 **M/r = 1.5***6/1 = 9 m/s

$\Delta H = -(a/g)\Delta v = -(1200/9.881)*9 = 1100.9$ m

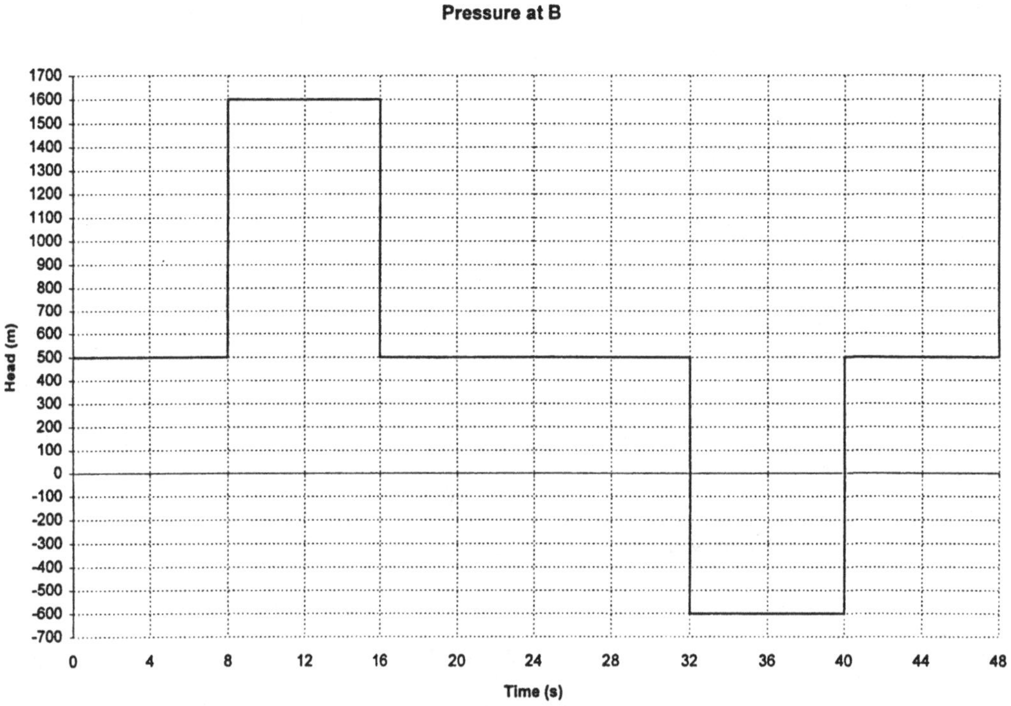

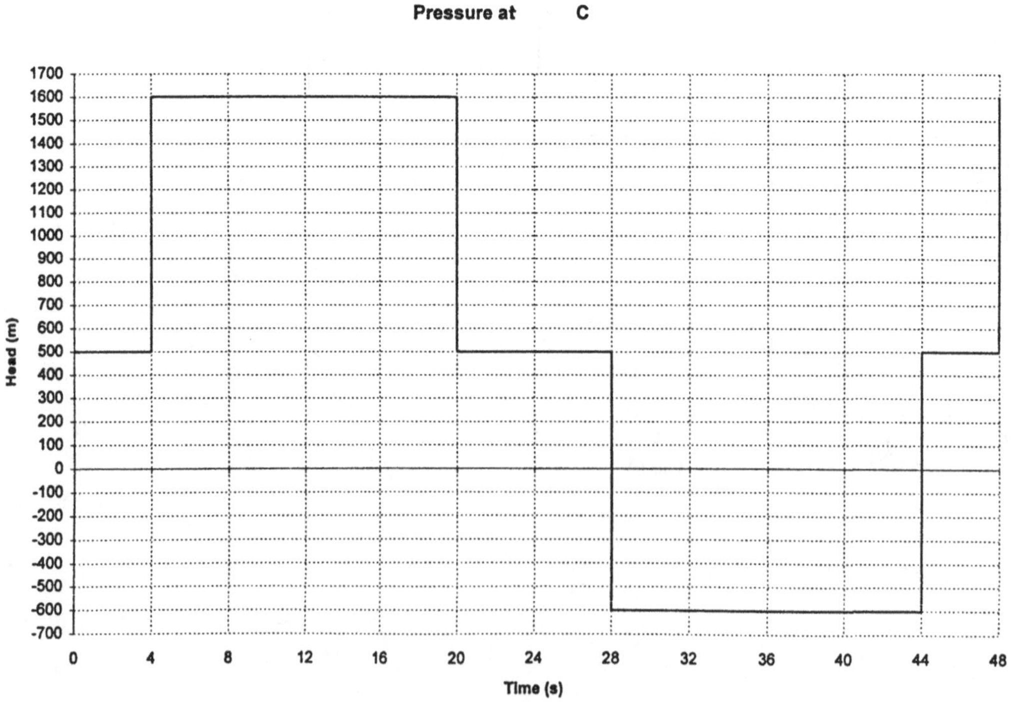

11-5

D = 1 m, e = 30 mm, E = 200 Gpa, K = 2.19 Gpa, ρ = 990 kg/m3, Vo = 1 m/s
Using Eq. 11-18

$$a = \sqrt{\left(\frac{2.19*10^9}{990}\right)/\left(1+\frac{2.19*1}{20*10^{-3}*200}\right)} = 1195.6 m/s$$

ΔH = -(a/g)Δv = - (1195.6/9.81)*(0 – 1) = 121.9 m

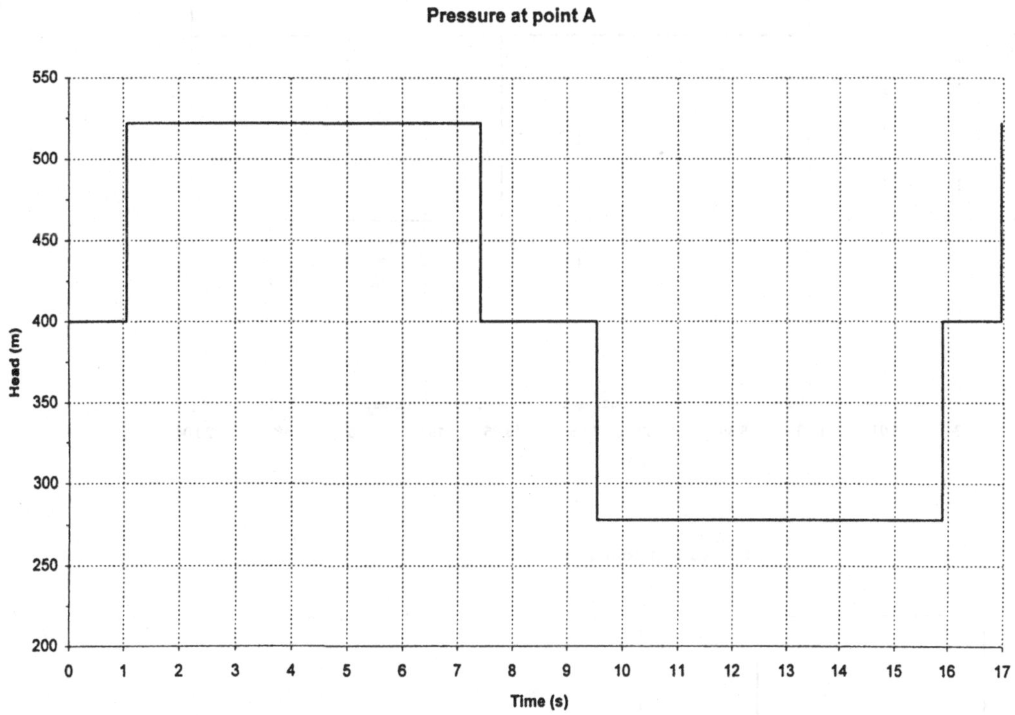

11-6

Take
L = 1000 m, a = 200 m/s
$\Delta H = -(a/g)\Delta v = -(200/9.81)*(0 - 0.981) = 20$ m

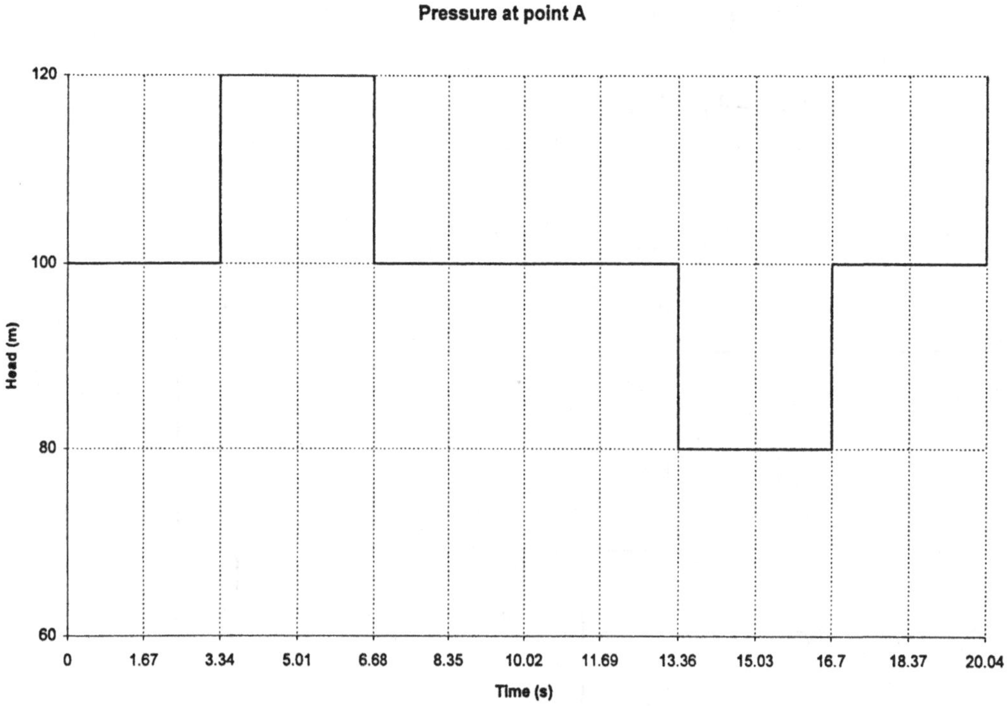

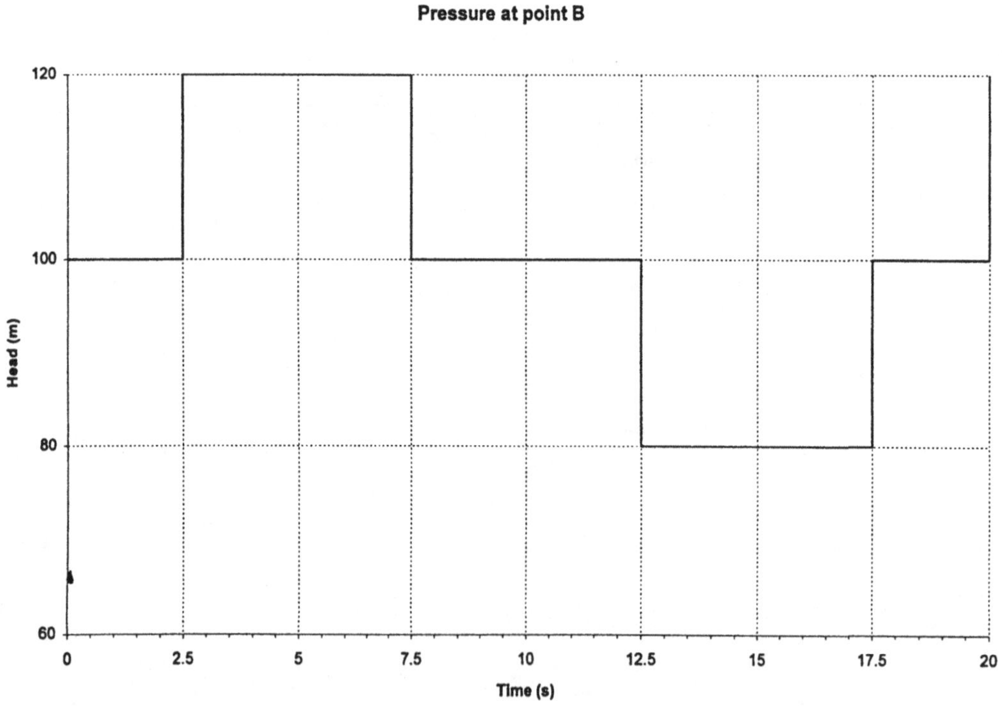

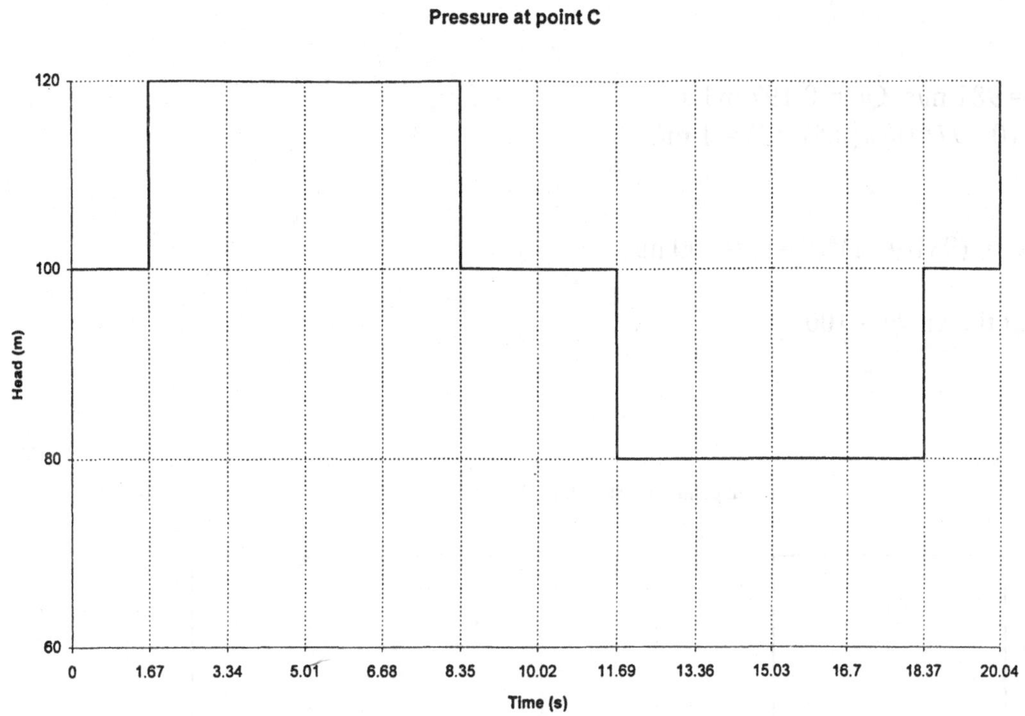

11-7

$\Delta v = 0 - 2 = -2$ ft/s, $a = 3220$ ft/s

$\Delta H = -(a/g)\Delta v = -(3220/32.2)*(-2) = 200$ ft

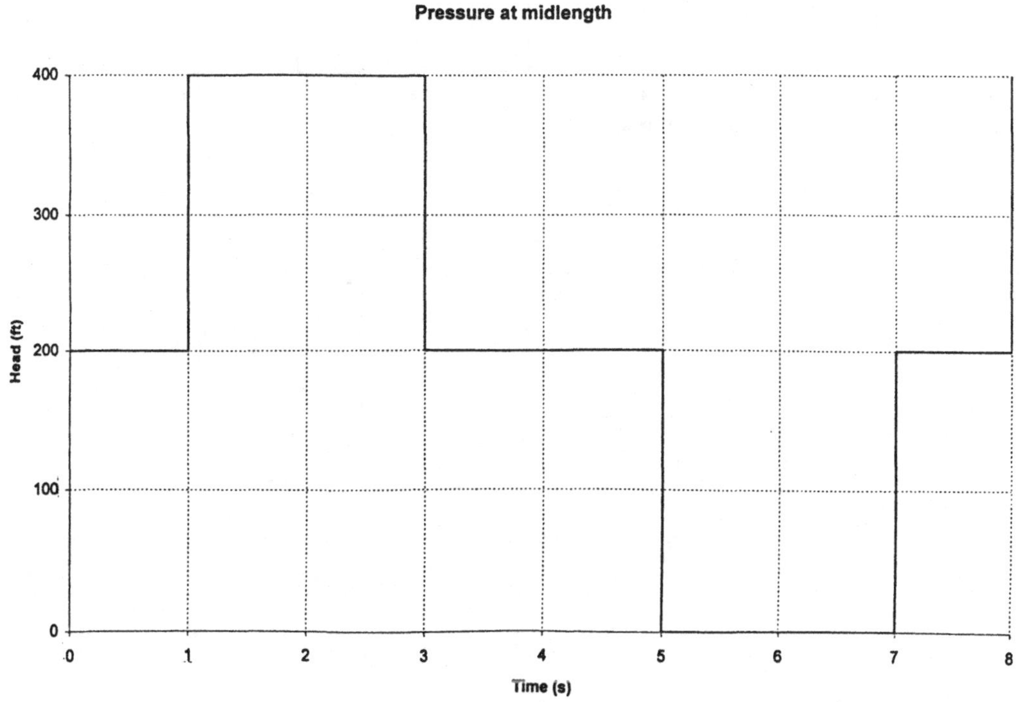

11-8

D = 0.5 m, a = 981 m/s, Qo = 0.197 m3/s
Vo = Qo/A = (0.197*4)/(π(0.5)**2) = 1 m/s

a)
ΔH = -(a/g)Δv = -(981/9.81)*(0 –1) = 100 m/s

Pressure rise at the valve = 100 m

b)

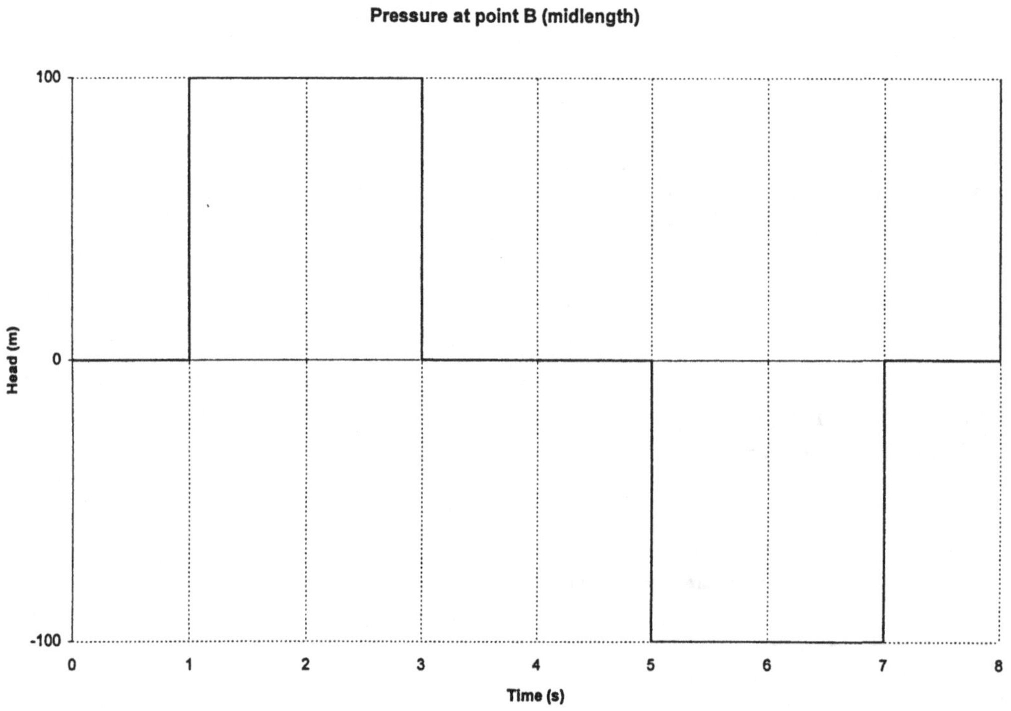

11-9

$\Delta H = -(a/g)\Delta v = -(3220)/32.2* (0 - 1) = 100$ ft

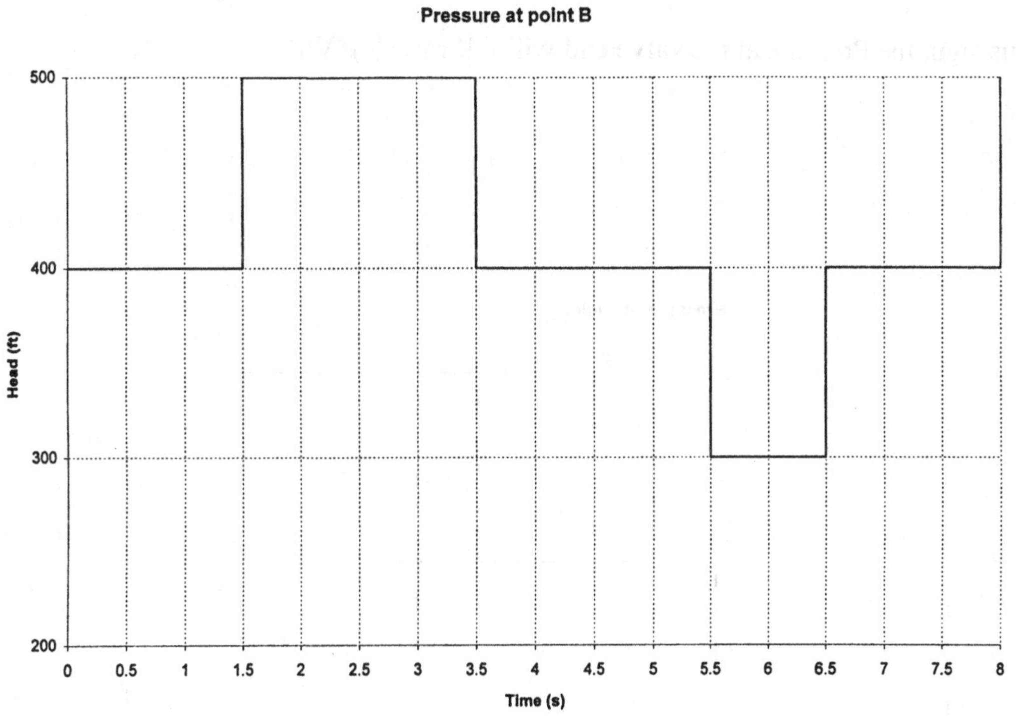

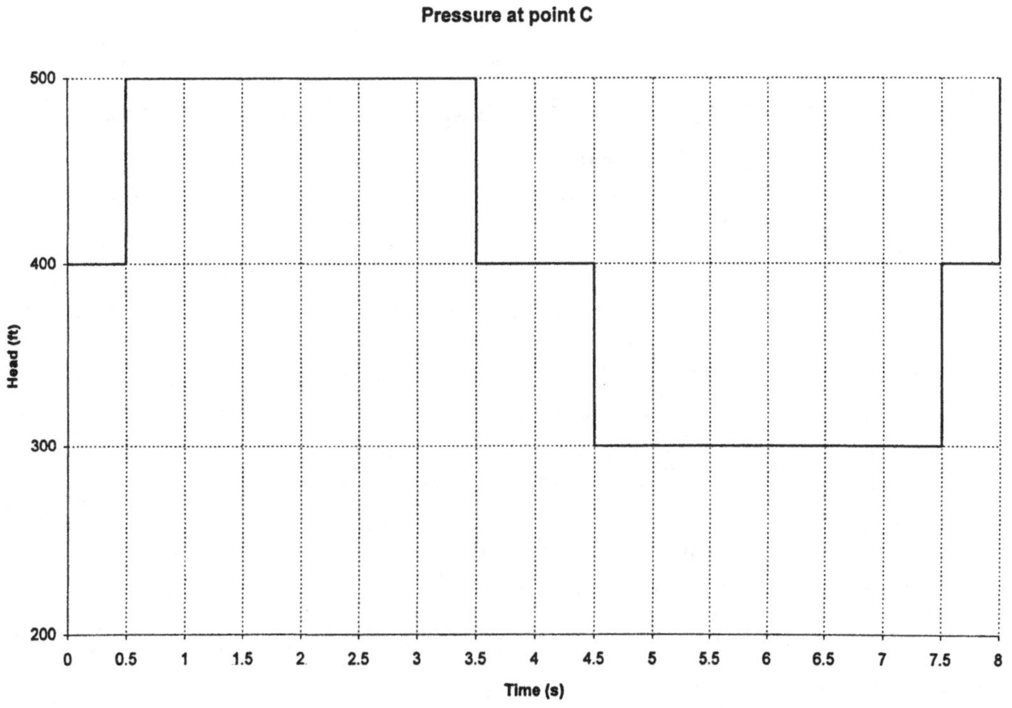

11-10

a)

$\Delta H = -(a/g)\Delta v$ where, $\Delta v = Vo - 0 = Vo$

$\Delta H = -(a/g)*Vo$

With a minus sign, the Pressure at the valve end will fall by $(a/g)*Vo$

b)

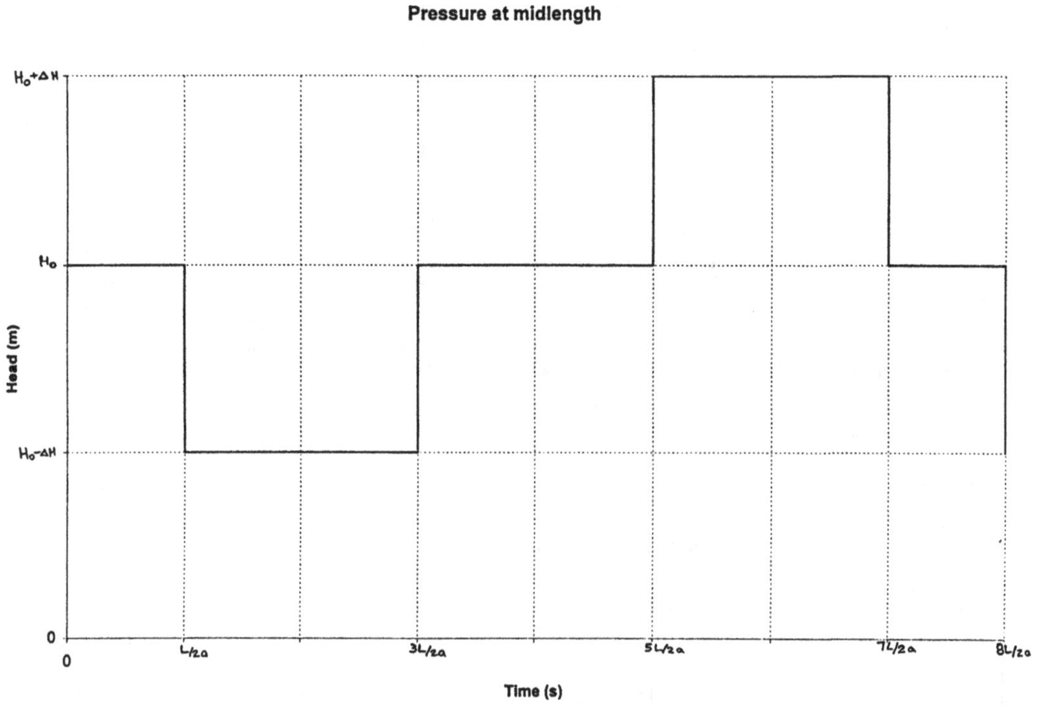

11-11

D = 1 m, a = 981 m/s, Vo = 2 m/s
ΔH = -(a/g)Δv = - (981/9.81)*(0 – 2) = 200 m

Pressure at 981 m upstream of valve (midlength)

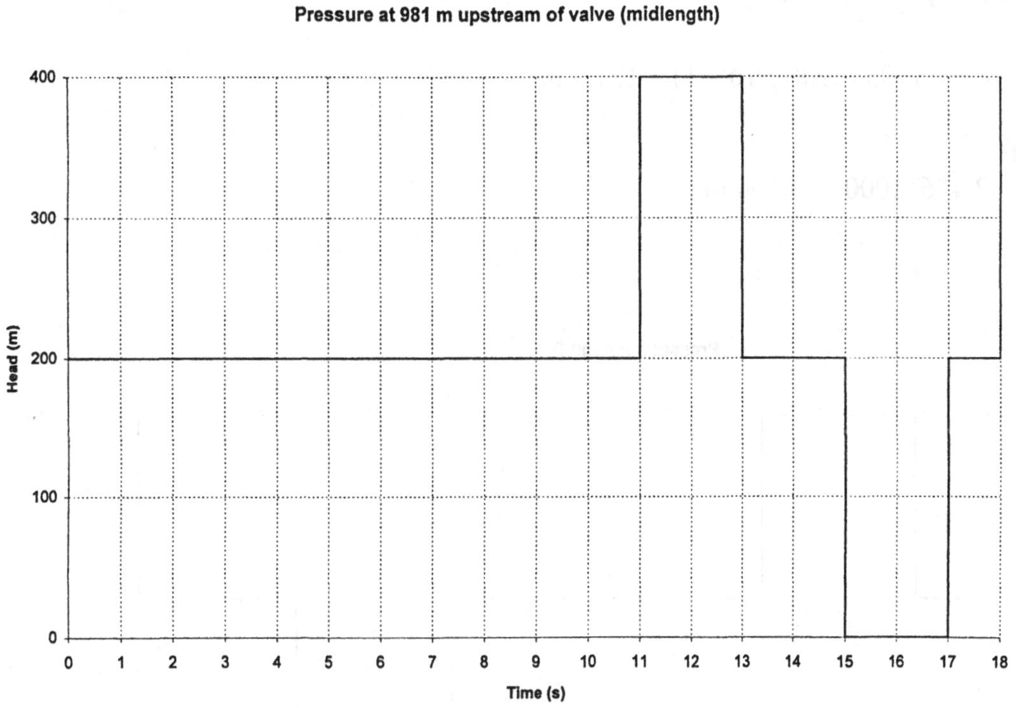

11-12

Vo = 1 m/s, D = 1 m, e = 30 mm, ρ = 990 kg/m3, K = 2.19 Gpa, E = 200 Gpa

$$a = \sqrt{\left(\frac{2.19*10^9}{990}\right) / \left(1 + \frac{2.19*1}{20*10^{-3}*200}\right)} = 1195.6 m/s$$

ΔH = -(a/g)Δv = -(1195.6/9.81)*(0 – 1) = 121.9 m

Take M = 6
L = 2.4*M = 2.4*6*1000 = 14400 m

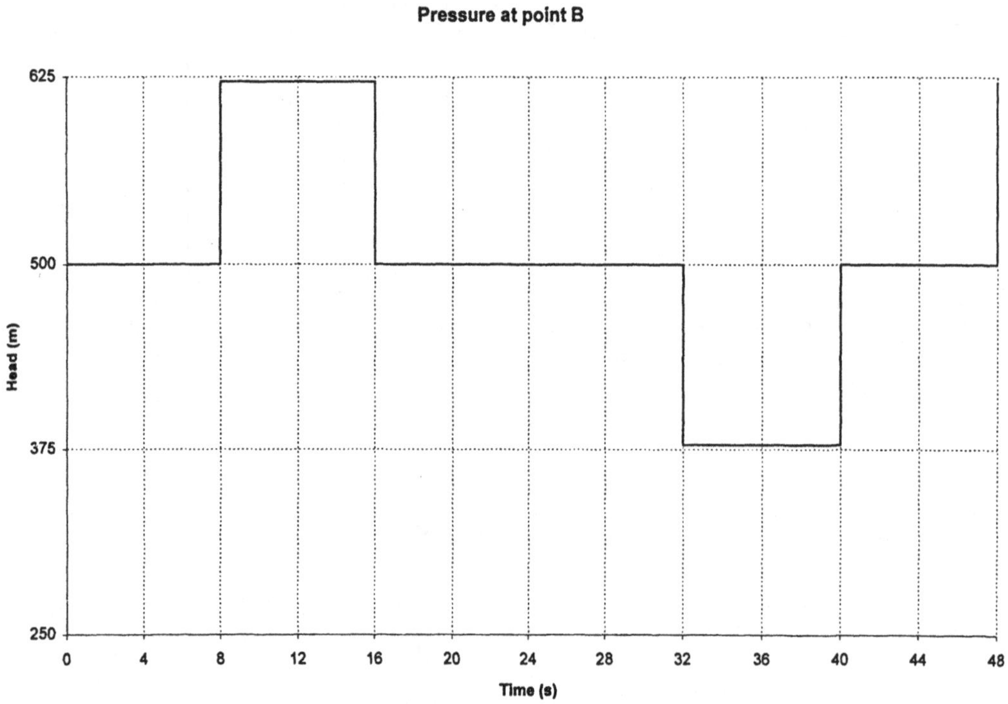

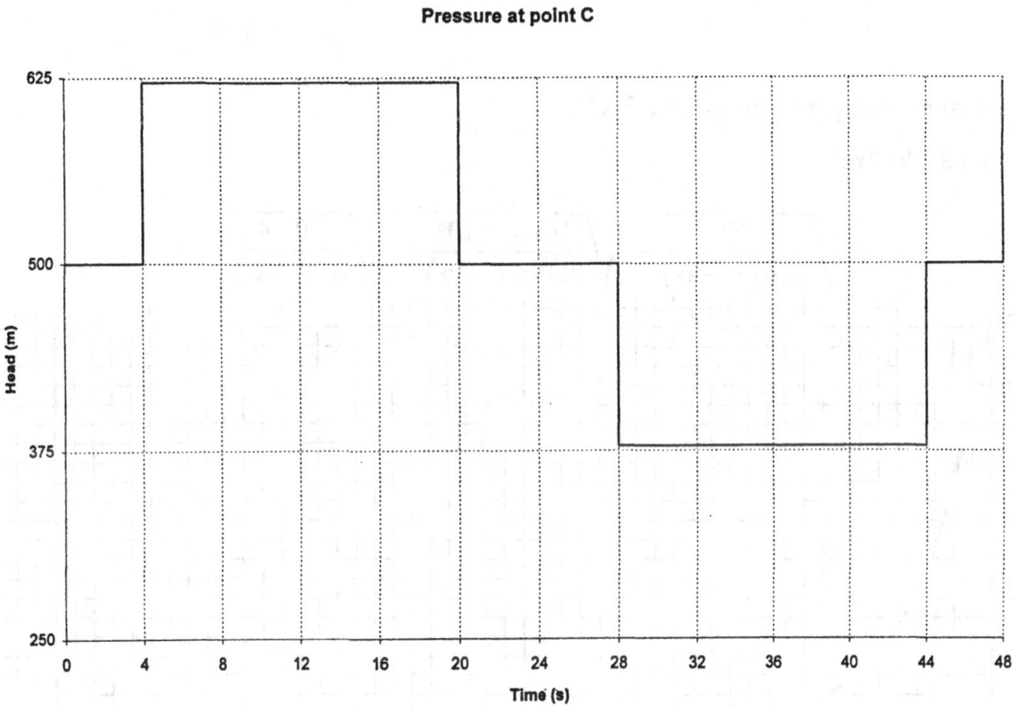

11-13

$\Delta H = -(a/g)\Delta v = -(3220/32.2)*(0 - 2) = 200$ ft

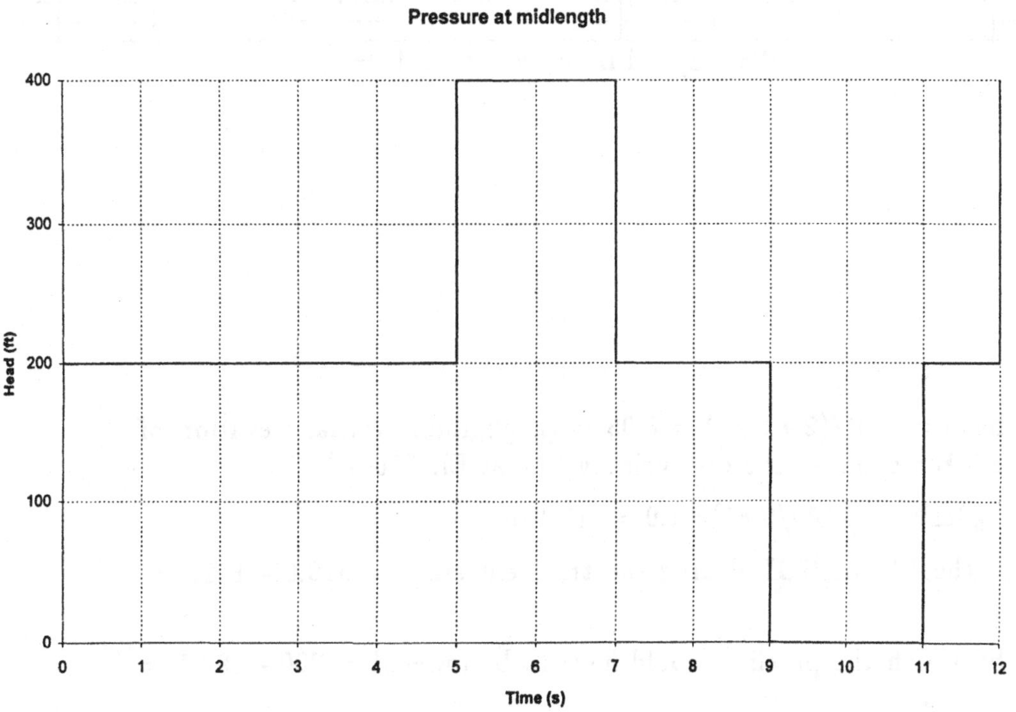

11-14

P^{*} = absolute pressure = 101.3 kPa

$\rho_L = 999\ kg/m^3$

$$a = \sqrt{\frac{P^{*}}{\rho_L \alpha(1-\alpha)}} = \sqrt{\frac{101.3 * 10^3}{999\alpha(1-\alpha)}} = \sqrt{\frac{101.4}{\alpha(1-\alpha)}}$$

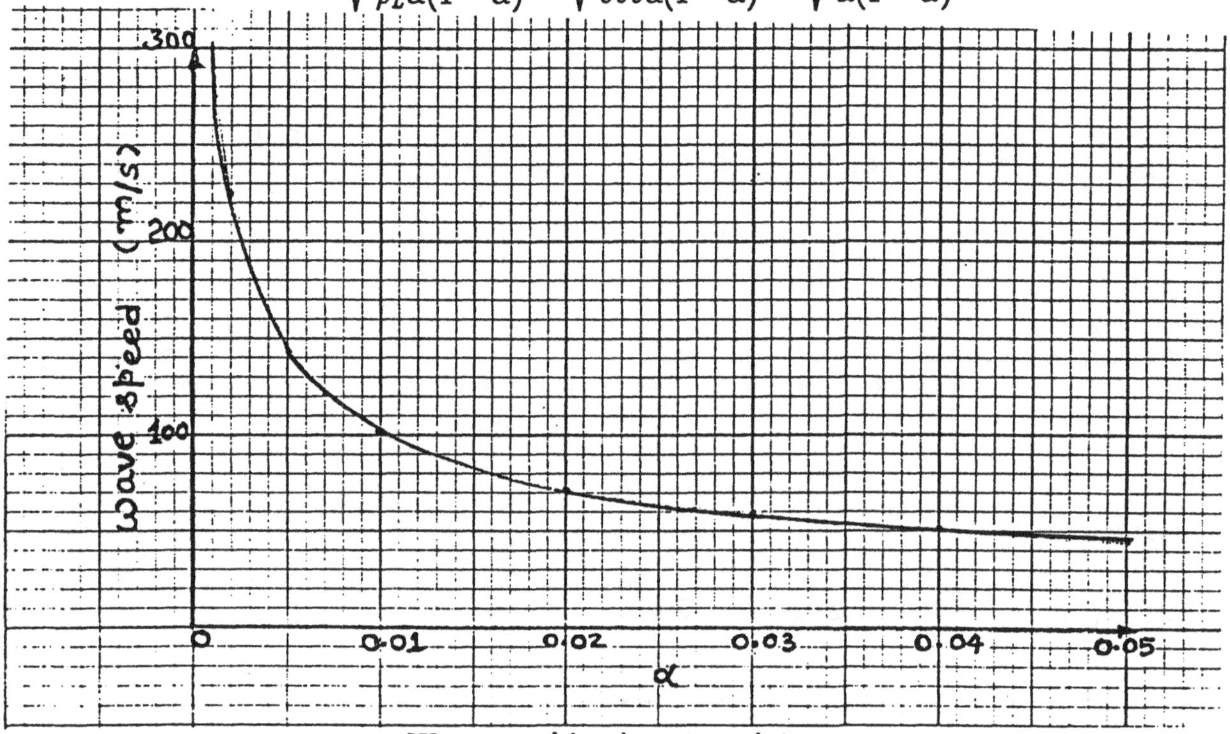

Wave speed in air-water mixture

11-15

Velocity head = $(1)^2/(2 * 9.81) = 0.05$ m (negligible). Initial elevation of HGL can be taken equal to the reservoir level i.e at El. 310 m.

$\Delta H = (a/g)\Delta V = (1100/9.81) * 1.0 = 112.1$ m

Therefore, the El. of HGL during the transient state = 310.0 - 112.1 = 197.1 m.

Amount by which the pipeline would have to be lowered = 200 - 197.1 = 2.9 m.

11-16

Program: Same as Fig. 11-16 of the text.

Input file

```
8,4,0.393,150.0,40.0
4000.0,0.5,1000.0,0.012
10.0,1.0,0.0,130.43,0.393
```

Output file

```
N =  8  IPRINT = 4  QO = 0.393 M3/S  HRES = 150.00 M  TLAST =  40.0 S

L =  4000.0 M  D =  0.50 M   A = 1000.0 M/S
F = 0.012

TV = 10.00  TAUO = 1.000  TAUF =0.000  HS = 130.43 M  QS = 0.393 M3/S

T =   0.00 S  TAU =1.000
 H =  150.00   147.55   145.10   142.65   140.20   137.75   135.30   132.85   130.40
 Q =   0.393    0.393    0.393    0.393    0.393    0.393    0.393    0.393    0.393

T =   2.00 S  TAU =0.800
 H =  150.00   147.55   145.10   142.65   140.20   143.44   147.12   151.36   156.13
 Q =   0.393    0.393    0.393    0.393    0.393    0.382    0.370    0.358    0.344

T =   4.00 S  TAU =0.600
 H =  150.00   152.97   156.37   160.33   164.79   169.87   175.52   181.86   188.83
 Q =   0.393    0.383    0.371    0.359    0.346    0.332    0.317    0.301    0.284

T =   6.00 S  TAU =0.400
 H =  150.00   161.39   172.84   184.45   196.28   203.67   211.75   220.64   230.27
 Q =   0.303    0.302    0.300    0.295    0.288    0.270    0.251    0.231    0.209

T =   8.00 S  TAU =0.200
 H =  150.00   165.77   181.62   197.62   213.84   230.38   247.25   264.57   282.34
 Q =   0.191    0.190    0.186    0.180    0.172    0.161    0.148    0.133    0.116

T =  10.00 S  TAU =0.000
 H =  150.00   171.16   192.38   213.76   235.31   252.89   269.94   286.45   302.31
 Q =   0.047    0.045    0.041    0.034    0.024    0.020    0.014    0.007    0.000

T =  12.00 S  TAU =0.000
 H =  150.00   173.31   195.84   217.52   238.29   242.51   245.51   247.28   247.88
 Q =  -0.140   -0.133   -0.129   -0.125   -0.123   -0.092   -0.061   -0.030    0.000

T =  14.00 S  TAU =0.000
 H =  150.00   154.22   157.99   161.09   163.53   168.53   172.12   174.20   174.93
 Q =  -0.288   -0.259   -0.227   -0.195   -0.162   -0.121   -0.081   -0.040    0.000

T =  16.00 S  TAU =0.000
 H =  150.00   135.02   119.98   104.75    89.37    88.07    86.25    83.59    80.33
 Q =  -0.183   -0.182   -0.178   -0.172   -0.164   -0.125   -0.085   -0.043    0.000

T =  18.00 S  TAU =0.000
 H =  150.00   129.65   109.28    88.76    68.20    51.43    35.34    19.75     4.99
 Q =  -0.045   -0.044   -0.040   -0.033   -0.023   -0.019   -0.013   -0.007    0.000
```

```
T =   24.00 S   TAU =0.000
  H =   150.00   164.27   178.58   193.09   207.72   209.23   211.12   213.81   216.95
  Q =    0.176    0.175    0.171    0.165    0.157    0.120    0.081    0.041    0.000

T =   26.00 S   TAU =0.000
  H =   150.00   169.60   189.19   208.92   228.57   244.60   259.82   274.59   288.37
  Q =    0.044    0.042    0.038    0.031    0.022    0.018    0.013    0.007    0.000

T =   28.00 S   TAU =0.000
  H =   150.00   171.36   191.96   211.77   230.49   234.78   237.71   239.58   240.10
  Q =   -0.129   -0.123   -0.118   -0.114   -0.111   -0.083   -0.055   -0.028    0.000

T =   30.00 S   TAU =0.000
  H =   150.00   153.74   157.16   160.04   162.34   167.05   170.46   172.47   173.16
  Q =   -0.262   -0.236   -0.208   -0.180   -0.149   -0.112   -0.075   -0.037    0.000

T =   32.00 S   TAU =0.000
  H =   150.00   136.39   122.75   108.88    94.93    93.25    91.29    88.58    85.56
  Q =   -0.169   -0.168   -0.164   -0.158   -0.150   -0.115   -0.078   -0.040    0.000

T =   34.00 S   TAU =0.000
  H =   150.00   131.09   112.23    93.23    74.41    59.06    44.63    30.61    17.69
  Q =   -0.042   -0.041   -0.037   -0.030   -0.021   -0.017   -0.012   -0.006    0.000

T =   36.00 S   TAU =0.000
  H =   150.00   129.49   109.74    90.76    72.91    68.61    65.72    63.82    63.34
  Q =    0.124    0.118    0.113    0.109    0.106    0.079    0.053    0.026    0.000

T =   38.00 S   TAU =0.000
  H =   150.00   146.46   143.20   140.42   138.18   133.60   130.28   128.29   127.63
  Q =    0.251    0.227    0.200    0.173    0.144    0.108    0.072    0.036    0.000

T =   40.00 S   TAU =0.000
  H =   150.00   163.01   176.04   189.31   202.64   204.47   206.48   209.21   212.12
  Q =    0.163    0.162    0.158    0.152    0.144    0.111    0.075    0.038    0.000
```

11-17

$L = 4000$ m, $a = 1000$ m/s, $D = 0.5$ m, $f = 0.012$, $H_0 = 150$ m.

Flow becomes fully established when $V = 0.99V_0$

$$H_0 = (1 + fL/D)\frac{V_0^2}{2g}$$

$$150 = (1 + 0.012 * 4000/0.5)\frac{V_0^2}{2 * 9.81}$$

or $V_0 = 5.51$ m/s.

Using Eq. 11-4,

$$T = (\frac{LV_0}{2gH_0})ln(1.99/0.01) = 5.293 * 4000 * 5.51/(2 * 9.81 * 150) = 39.63$$

Input File

```
8,4,0.0,150.0,100.0
4000.0,0.5,1000.0,0.012
0.0,0.0,1.0,1.547,1.0819
```

Time of flow establishment = 39.63 s.

Time of flow establishment using the program = 90 s.

11-18

Flow through orifice can be written as

$$Q_P = (C_d A_0)\sqrt{2gH_P} \quad (1)$$

in which subscript P indicates the transient state conditions. Q = discharge, H = head upstream of the valve, C_d = coefficient of discharge and A_0 = area of orifice opening. At the downstream end, positive characteristic equation, Eq. 11-55 is valid.

$$Q_P = C_P - C_a H_P \quad (2)$$

Defining

$$2gC_d^2 A_0^2 / C_a = C_v \quad (3)$$

$$Q_P^2 = C_a C_v H_P \quad (4)$$

and solving Eqs. 1-4, following equation can be obtained for Q_p.

$$Q_P = 0.5\left(-C_v + \sqrt{C_v^2 + 4C_P C_v}\right)$$

Steady state flow through the valve can be written as

$$Q_0 = (C_d A_v)_0 \sqrt{2g\Delta H_0} \tag{1}$$

in which subscript 0 indicates steady state conditions, C_d = coefficient of discharge, ΔH = head differential, A_v = area of valve opening. A similar equation may be written for the transient state as follows.

$$Q_P = (C_d A_v)\sqrt{2g\Delta H_P} \tag{2}$$

Defining

$$\tau = \frac{C_d A_v}{(C_d A_v)_0} \tag{3}$$

$$C_v = \frac{(Q_0 \tau)^2}{\Delta H_0} \tag{4}$$

Eq. 2 can be written as

$$Q_P{}^2 = C_v \Delta H_P \tag{5}$$

Assuming flow in the downstream direction;

$$\Delta H_P = H_{i,n+1} - H_{i+1,1} \tag{6}$$

$$Q_P = Q_{i,n+1} = Q_{i+1,1} \tag{7}$$

negative characteristic:

$$Q_{i+1,1} = C_{n_{i+1}} + C_{a_{i+1}} H_{i+1,1} \tag{8}$$

positive characteristic:

$$Q_{i,n+1} = C_{p_i} - C_{a_i} H_{i,n+1} \tag{9}$$

Eqs. 5-9 can be reduced to a single equation,

$$\frac{C_{p_i} - Q_P}{C_{a_i}} = \frac{Q_P - C_{n_{i+1}}}{C_{a_{i+1}}} + \frac{Q_P{}^2}{C_v} \tag{10}$$

Above quadratic equation can be solved for Q_P.

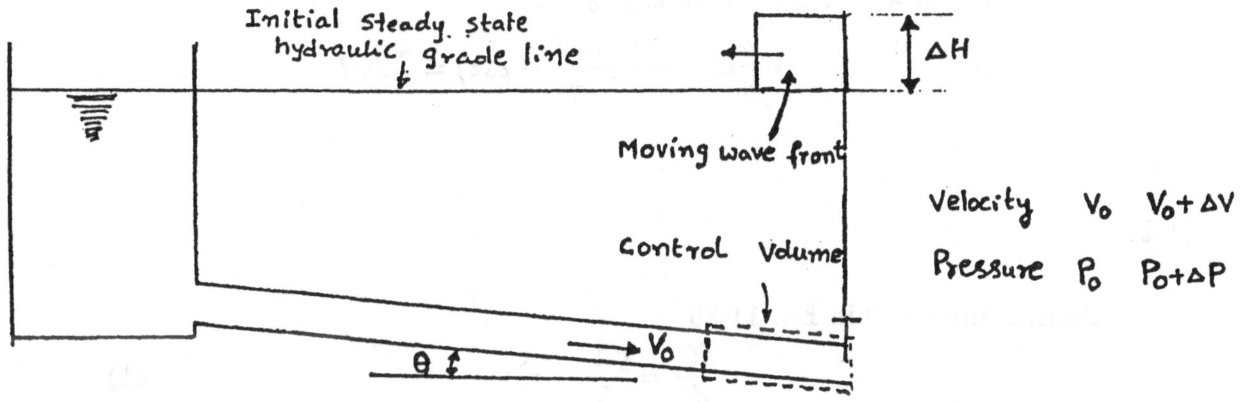

(a) unsteady flow

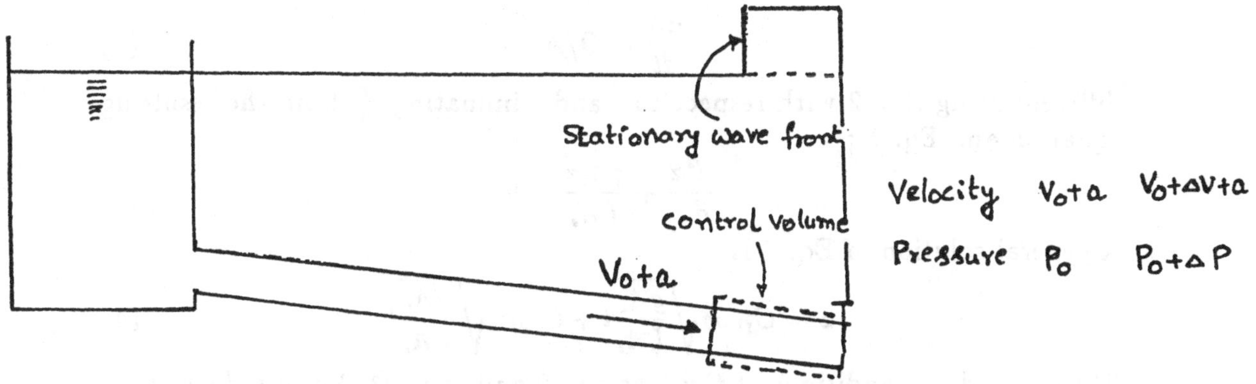

(b) Unsteady flow Converted to Steady flow

Control volumes for unsteady flow and unsteady flow converted to steady flow by superimposing wave velocity a are shown in the above figures.

Rate of change of momentum in positive x-direction

$$= \rho(V_o + a)A((V_o + \Delta V + a) - (V_o + a)) = \rho(V_o + a)A\Delta V$$

neglecting friction, resultant force acting on the fluid in the positive x-direction

$$= P_0 A - (P_o + \Delta P)A + A\Delta x \rho g sin\theta = -\Delta PA + A\Delta x \rho g sin\theta$$

applying Newtons second law

$$\rho(V_0 + a)A\Delta V = -\Delta PA + A\Delta x\rho g\, sin\theta$$

in which $sin\theta = -\frac{\Delta z}{\Delta x}$. Assuming $V_0 << a$,

$$\frac{a}{g}\Delta V = -(\frac{\Delta P}{\gamma} + \Delta z) = -\Delta H$$

11-21

Substituing $c = 0$ in Eq. 11-86

$$\frac{dQ}{dt} = \frac{gA_t}{L}(-z) \qquad (1)$$

$Q_{tur} = 0$ at $t \geq 0$

Eq. 11-88 can be written as

$$\frac{dZ}{dt} = Q/A_s \qquad (2)$$

differentiating Eq. 2 with respect to t and eliminating $\frac{dQ}{dt}$ from the resulting equation and Eq. 1 ;

$$\frac{d^2 z}{dt^2} + \frac{gA_t z}{LA_s} = 0 \qquad (3)$$

A general solution of Eq. 3 is

$$z = C_1 cos\sqrt{\frac{gA_t}{LA_s}}t + C_2 sin\sqrt{\frac{gA_t}{LA_s}}t \qquad (4)$$

With boundary conditions at $t = 0$ as $z = 0$ and $\frac{dz}{dt} = \frac{Q_0}{A_s}$, Eq. 4 reduces to

$$z = Q_0\sqrt{\frac{L}{gA_sA_t}}sin\sqrt{\frac{gA_t}{LA_s}}t \qquad (5)$$

Amplitude Z is obtained from the above equation by substituting

$$sin\sqrt{\frac{gA_t}{LA_s}}t = 1$$

Therefore,

$$Z = Q_0\sqrt{\frac{L}{gA_sA_t}}$$

Period T is obtained by solving

$$sin\sqrt{\frac{gA_t}{LA_s}}T = 0$$

or

$$T = 2\pi\sqrt{\frac{LA_s}{gA_t}}$$

12-1

First derive the relationship for $2S/\Delta t + O$. From the stated conditions of the reservoir $S = 1{,}000$ Acr. x 43,560 ft^2/Acr. x H ft where H is the head on the spillway. Let $\Delta t = 1$ hr = 3600 sec. Then

$$2S/\Delta t = 2 \times 43{,}560{,}000/3600 \text{ ft}^3/\text{sec}$$

$$= 24{,}200 \text{ H ft}^3/\text{sec}$$

$$Q = K \sqrt{2g}\, L\, H^{3/2}$$

$$= 0.50 \sqrt{2g}\,(200)\,(H^{3/2})$$

$$Q = 802.5\, H^{3/2} \text{ or } H = (Q/(802.5))^{2/3}$$

then $2S/\Delta t + O = 24{,}200\, H + 802.5\, H^{3/2}$

or $\quad 2S/\Delta t + O = 24{,}200\,(Q/802.5)^{2/3} + Q \qquad (1)$

Now develop the table of t vs O as in Example 4-11 on page 000. The inflow values are all given. To calculate the magnitude of $2S_i/\Delta t - O_i$ for the first entry of Column 5 we can calculate the initial value of $2S/\Delta t + O$ and subtract 20 as below:

$$(2S/\Delta t)_i - O_i = (2S/\Delta t)_i + O_i - 2O_i$$

or from (1), above,

$$(2S/\Delta t)_i - O_i = 24{,}200\,(Q/802.5)^{2/3} - Q$$

$$= 24{,}200\,(200/802.5)^{2/3} - 200$$

$$= 9384 \text{ cfs}$$

Then $(2S/\Delta t)_2 + O_2 = 9384 + I_1 + I_2 = 10{,}584$ cfs

or $\quad (2S/\Delta t)_2 + O_2 = 10{,}584 = 24{,}200\,(O_2/802.5)^{2/3} + O_2$

$$O_2 = 225 \text{ cfs}$$ ← iterate

The rest of the table is completed utilizing Eq (1) (above) to solve for O given $2S/\Delta t + O$. One could also make a plot of O vs $2S/\Delta t + O$ (obtained from Eq (1) and utilize that plot to develop the table.

(Table on separate page)

TABLE FOR PROBLEM 12-1

(1) t (hr)	(2) i	(3) I_i (cfs)	(4) I_i+I_{i+1} (cfs)	(5) $\frac{2Si}{t} - O_i$ (cfs)	(6) $\frac{2S_{i+1}}{t} + O_{i+1}$ (cfs)	(7) O_i (cfs)
0	1	200	1,200	9,384	10,584	200
1	2	1,000	4,000	10,134	14,134	225
2	3	3,000	8,000	13,438	21,438	348
3	4	5,000	13,000	20,160	33,160	639
4	5	8,000	20,000	30,726	50,726	1,217
5	6	12,000	30,000	46,178	76,178	2,274
6	7	18,000	41,000	67,932	108,932	4,123
7	8	23,000	47,000	95,044	142,044	6,944
8	9	24,000	46,000	121,644	167,444	10,200
9	10	22,000	42,000	141,544	183,544	12,950
10	11	20,000	39,000	153,984	192,984	14,780
11	12	19,000	33,000	161,214	194,214	15,885
12	13	14,000	24,000	162,144	186,144	16,035
13	14	10,000	17,000	155,984	172,984	15,080
14	15	7,000	12,000	145,844	157,844	13,570
15	16	5,000	9,000	134,064	143,064	11,890
16	17	4,000	7,000	122,444	129,444	10,310
17	18	3,000	5,000	111,604	116,604	8,920
18	19	2,000	3,500	101,284	104,784	7,660
19	20	1,500	2,500	91,664	94,164	6,560
20	21	1,000	1,500	82,934	84,434	5,615
21	22	500	900	74,854	75,754	4,790
22	23	400	700	67,574	68,274	4,090
23	24	300				3,480

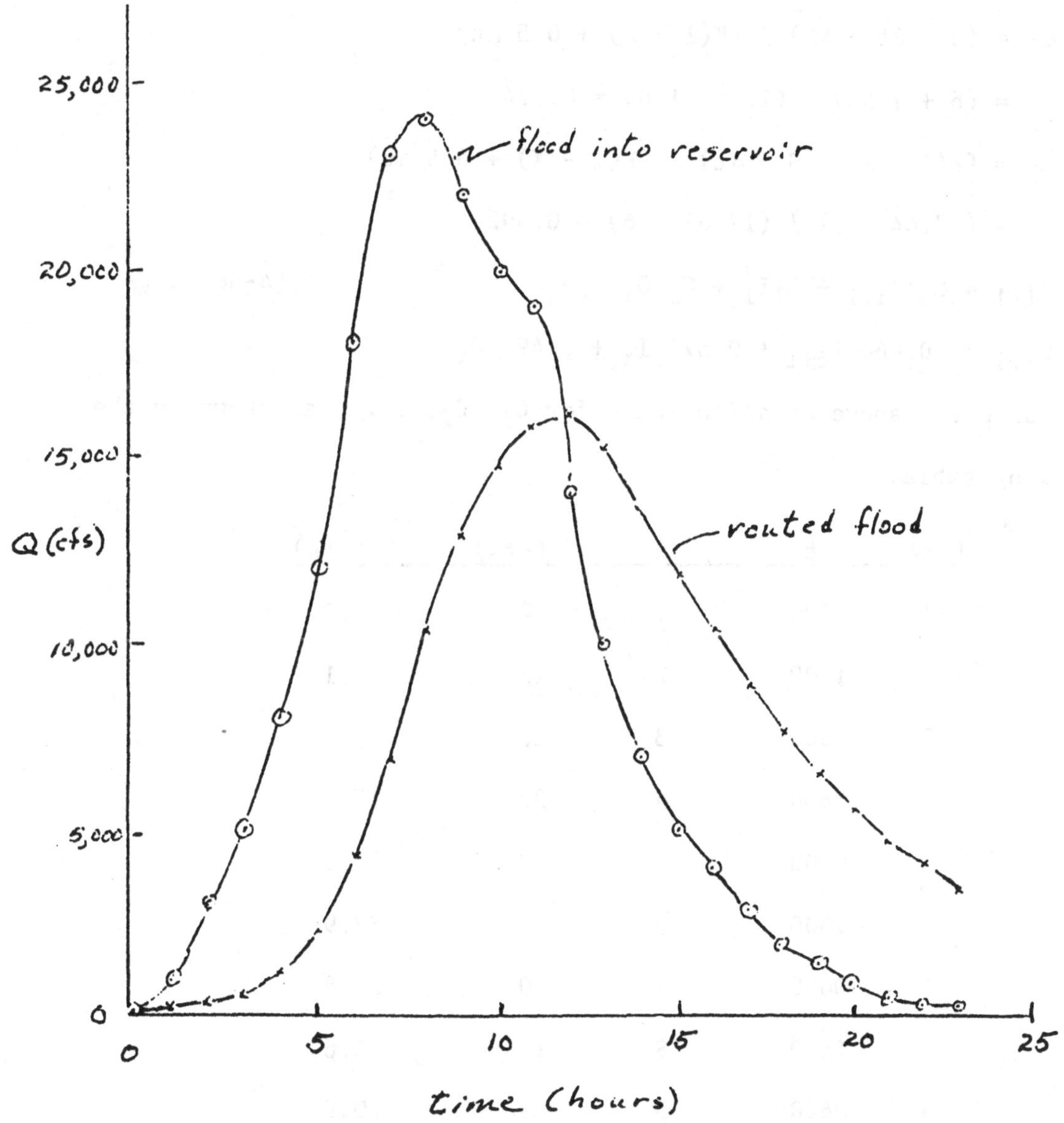

HYDROGRAPHS FOR PROBLEM 12-1

12-2

First solve for C_o, C_1 and C_2.

$C_o = (0.5\ \Delta t - KX)\ /\ (K(1 - X) + 0.5\ \Delta t)$

where $K(1 - X) = 25.2\ (1 - 0.3) = 17.64$ hrs

$0.5\ \Delta t = 0.5 \times 12 = 6$ hrs

$KX = 25.2 \times 0.3 = 7.56$ hrs.

Then $C_o = (6 - 7.56) / (17.64 + 6) = -0.0660$

$C_1 = (0.5\ \Delta t + KX) / (K(1 - X) + 0.5\ \Delta t)$

$= (6 + 7.56) / (17.64 + 6) = 0.574$

$C_2 = (K(1 - X) - 0.5\ \Delta t) / (K(1 - X) + 0.5\ \Delta t)$

$= (17.64 - 6) / (17.64 + 6) = 0.492$

Then $O_{i+1} = C_o I_{i+1} + C_1 I_i + C_2 O_i$ (4-56)

$O_{i+1} = -0.066\ I_{i+1} + 0.574\ I_i + 0.492\ O_i$

Next using the above equation solve for O_1, O_2, . . . as shown in the following table.

Date	t	i	I (cfs)	O (cfs)
1	0600	1	2	2.0
	1800	2	15	1.1
2	0600	3	28	7.3
	1800	4	32	17.6
3	0600	5	30	25.0
	1800	6	25	27.9
4	0600	7	20	26.8
	1800	8	16	23.6
5	0600	9	13	19.9
	1800	10	10	16.6
6	0600	11	7	13.4
	1800	12	5	10.3
7	0600	13	4	7.7
	1800	14	3	5.9
8	0600	15	3	4.4
	1800	16	2	3.8

(i) Constant-level upstream reservoir

Neglecting velocity head and entrance losses,

$$Y_1^{k+1} = Y_{res} \tag{1}$$

in which Y_{res} = flow depth in the reservoir above the channel bottom at node 1. Negative characteristic equation (Eq. 12.37) is applicable at the upstream end.

$$V_1^{k+1} = C_n + K_2 Y_1^{k+1} \tag{2}$$

in which

$$C_n = V_2^k - K_2 Y_2^k + g(S_0 - S_f)_2 \Delta t \tag{3}$$

and

$$K_2 = g/c_2 \tag{4}$$

Substituting Eqs. 1,3 and 4 in Eq. 2

$$V_1^{k+1} = V_2^k - g/c_2 Y_2^k + g(S_0 - S_f)_2 \Delta t + (g/c_2) Y_{res}$$

(ii) Constant-level downstream reservoir

$$Y_{N+1}{}^{k+1} = Y_{res} \tag{5}$$

Positive characteristic equation is applicable at the downstream end.

$$V_{N+1}{}^{k+1} = C_p - K_N Y_{N+1}{}^{k+1} \tag{6}$$

in which

$$C_p = V_N{}^k + K_N Y_N{}^k + g(S_0 - S_f)_N \Delta t \tag{7}$$

and

$$K_N = g/c_N \tag{8}$$

Substituting Eqs. 5,7 and 8 in Eq. 6

$$V_{N+1}{}^{k+1} = V_N{}^k + (g/c_N) Y_N{}^k + g(S_0 - S_f)_N \Delta t - (g/c_N) Y_{res}$$

(iii) Rating curve for the downstream end

From positive characteristic equation,

$$V_{N+1}{}^{k+1} = V_N{}^k + (g/c_N) Y_N{}^k + g(S_0 - S_f)_N \Delta t - (g/c_N) Y_{N+1}{}^{k+1} \tag{9}$$

Rating curve gives,

$$Q = f(Y) \tag{10}$$

in which f stands for a function. Writing Eq. 2 in terms of velocity

$$V = f_1(Y) \tag{11}$$

or

$$V_{N+1}{}^{k+1} = f_1(Y_{N+1}{}^{k+1}) \tag{12}$$

Equations 9 and 12 give two equations in two unknowns $V_{N+1_{k+1}}$ and $Y_{N+1}{}^{k+1}$. They can be solved by Newton-Raphson technique.

12-4

```
C
C       COMPUTAT ON OF UNSTEADY, FREE-SURFACE FLOWS BY LAX'S
C        DIFFUSIVE SCHEME
C
C       ******************** NOTATION ***********************
C
C       A = FLOW AREA;
C       B = TOP WATER-SURFACE WIDTH;
C       BO = CHANNEL-BOTTOM WIDTH;
C       P = WETTED PERIMETER;
C       Q = DISCHARGE;
C       HR = HYDRAULIC RADIUS;
C       S = CHANNEL-SIDE SLOPE, S HORIZONTAL : 1 VERTICAL;
C       SO = CHANNEL-BOTTOM SLOPE;
C       X = DISTANCE ALONG CHANNEL BOTTOM, POSITIVE IN THE
C           DOWNSTREAM DIRECTION;
C       Y = FLOW DEPTH;
C       YRES = DEPTH OF FLOW AT LAST NODE;
C       VU = VELOCITY AT UPSTREAM END.
C
        REAL L, MN,MN2
        DIMENSION Y(200),YP(200),V(200),VP(200)
        BT(YY)=BO+2.*S*YY
        AR(YY)=YY*(BO+S*YY)
        WP(YY)=BO+2.*YY*SQRT(1.+S*S)
        G=9.81
        READ (5,*) N,IPRINT,YD,MN,BO,SO,S,L,TMAX
        WRITE(6,10) N,YD,MN,BO,SO,S,L
 10     FORMAT(5X,'N =',I3,' YD =',F6.2,' M',
     1 2X,' MN =',F6.3,' BO =',F6.2,' M'/5X,' SO =',F6.4,'S =',F8.4,
     2' L =',F8.2,' M')
C
C      STEADY-STATE CONDITIONS
C
       MN2=MN*MN
       NN=N+1
       A=AR(YD)
       P=WP(YD)
       HR=A/P
       VO=HR**0.6667*SQRT(SO)/MN
       DO 30 I = 1,NN
       Y(I) = YD
       V(I)=VO
 30    CONTINUE
       YRES=Y(NN)
       VU= 0.0
       B=BT(YD)
       C=SQRT(G*A/B)
       DX=L/N
       DT=DX/(VO+C)
       T=0.0
 35    K = 0
       WRITE(6,40) T
 40    FORMAT(/5X,'T =',F8.3,' S')
       WRITE(6,50) (Y(I),I=1,NN)
```

```
50    FORMAT(6X,'Y =',(11F6.2))
      WRITE (6,60) (V(I),I=1,NN)
60    FORMAT(6X,' V=',(12F6.2))
70    T=T+DT
      K=K+1
      R=0.5*DT/DX
      IF (T.GT.TMAX) GO TO 160
C
C     UPSTREAM END
C
      VP(1) = VU
      AB=AR(Y(2))
      BB=BT(Y(2))
      CB=SQRT(G*BB/AB)
      RB=AB/WP(Y(2))
      SFB=(MN2*V(2)*V(2))/(RB**1.333)
      CN=V(2)-CB*Y(2)+G*(SO-SFB)*DT
      YP(1)=(VP(1)-CN)/CB
C
C     DOWNSTREAM END
C
      YP(NN)=YRES
      AA=AR(Y(N))
      BA=BT(Y(N))
      CA=SQRT(G*BA/AA)
      RA=AA/WP(Y(N))
      SFA=(MN2*V(N)*V(N))/(RA**1.333)
      CP=V(N)+CA*Y(N)+G*(SO-SFA)*DT
      VP(NN)=CP-CA*YP(NN)
C
C     INTERIOR NODES
C
      DO 80 I=2,N
      I1=I-1
      IP1=I+1
      AA=AR(Y(I1))
      PA=WP(Y(I1))
      RA=AA/PA
      SFA=(MN2*V(I1)*V(I1))/(RA**1.333)
      BA=BT(Y(I1))
      AB=AR(Y(IP1))
      BB=BT(Y(IP1))
      P=WP(Y(IP1))
      RB=AB/P
      SFB=(MN2*V(IP1)*V(IP1))/(RB**1.333)
      DM=0.5*(AA/BA + AB/BB)
      SFM=0.5*(SFA+SFB)
      VM=0.5*(V(I1)+V(IP1))
      YM=0.5*(Y(I1)+Y(IP1))
      VP(I)=VM-R*G*(Y(IP1)-Y(I1)) - R*VM*(V(IP1)-V(I1))
     1    + G*DT*(SO-SFM)
      YP(I)=YM-R*DM*(V(IP1)-V(I1))-R*VM*(Y(IP1)-Y(I1))
80    CONTINUE
C
C     CHECK FOR STABILITY

C
      DO 100 I=1,NN
      A=AR(YP(I))
      B=BT(YP(I))
      C=SQRT(G*A/B)
      DTN=DX/(ABS(VP(I))+C)
      IF (DTN.LE.DT) GO TO 110
      IF (DT.LT.0.75*DTN) DTNEW=1.15*DT
      IF (DTN.GE.0.75*DT) DTNEW=DT
100   CONTINUE
      GO TO 120
C
C     REDUCE DT FOR STABILITY AND RE-CALCULATE
C
110   T=T-DT
      DT=.9*DTN
      K=K-1
      GO TO 70
120   DT=DTNEW
      DO 130 I=1,NN
      V(I)=VP(I)
      Y(I)=YP(I)
130   CONTINUE
      IF (K.EQ.IPRINT) GO TO 35
      GO TO 70
160   STOP
      END
```

Input file

```
10,1,3.0,0.025,10.0,0.0001,1.5,5000.0,2000.0
```

Output file

```
N = 10 YD =   3.00 M    MN = 0.025 BO = 10.00 M
 SO =0.0001S =   1.5000 L = 5000.00 M

T =    0.000 S
 Y =  3.00  3.00  3.00  3.00  3.00  3.00  3.00  3.00  3.00  3.00  3.00
  V=  0.65  0.65  0.65  0.65  0.65  0.65  0.65  0.65  0.65  0.65  0.65

T =   92.713 S
 Y =  2.68  3.00  3.00  3.00  3.00  3.00  3.00  3.00  3.00  3.00  3.00
  V=  0.00  0.65  0.65  0.65  0.65  0.65  0.65  0.65  0.65  0.65  0.65

T =  185.426 S
 Y =  2.68  2.70  3.00  3.00  3.00  3.00  3.00  3.00  3.00  3.00  3.00
  V=  0.00  0.07  0.65  0.65  0.65  0.65  0.65  0.65  0.65  0.65  0.65

T =  278.139 S
 Y =  2.63  2.70  2.72  3.00  3.00  3.00  3.00  3.00  3.00  3.00  3.00
  V=  0.00  0.07  0.11  0.65  0.65  0.65  0.65  0.65  0.65  0.65  0.65

T =  370.852 S
 Y =  2.63  2.65  2.72  2.74  3.00  3.00  3.00  3.00  3.00  3.00  3.00
  V=  0.00  0.06  0.11  0.15  0.65  0.65  0.65  0.65  0.65  0.65  0.65

T =  463.565 S
 Y =  2.58  2.65  2.68  2.74  2.76  3.00  3.00  3.00  3.00  3.00  3.00
  V=  0.00  0.06  0.11  0.15  0.19  0.65  0.65  0.65  0.65  0.65  0.65

T =  556.278 S
 Y =  2.58  2.61  2.68  2.70  2.76  2.77  3.00  3.00  3.00  3.00  3.00
  V=  0.00  0.06  0.11  0.16  0.19  0.23  0.65  0.65  0.65  0.65  0.65

T =  648.990 S
 Y =  2.54  2.61  2.63  2.70  2.72  2.77  2.79  3.00  3.00  3.00  3.00
  V=  0.00  0.06  0.11  0.16  0.21  0.23  0.26  0.65  0.65  0.65  0.65

T =  741.703 S
 Y =  2.54  2.56  2.63  2.66  2.72  2.75  2.79  2.80  3.00  3.00  3.00
  V=  0.00  0.06  0.11  0.16  0.21  0.25  0.26  0.29  0.65  0.65  0.65

T =  834.416 S
 Y =  2.50  2.56  2.59  2.66  2.68  2.75  2.77  2.80  2.82  3.00  3.00
  V=  0.00  0.06  0.11  0.16  0.21  0.25  0.29  0.29  0.31  0.65  0.65

T =  927.129 S
 Y =  2.50  2.52  2.59  2.62  2.68  2.71  2.77  2.79  2.82  2.83  3.00
  V=  0.00  0.06  0.11  0.16  0.21  0.25  0.29  0.32  0.31  0.34  0.65

T =1019.842 S
 Y =  2.46  2.52  2.55  2.62  2.64  2.71  2.73  2.79  2.80  2.83  3.00
  V=  0.00  0.06  0.11  0.16  0.20  0.25  0.28  0.32  0.36  0.34  0.04

T =1112.555 S
 Y =  2.46  2.48  2.55  2.58  2.64  2.67  2.73  2.75  2.80  2.96  3.00
  V=  0.00  0.06  0.11  0.15  0.20  0.24  0.28  0.32  0.36  0.10  0.04
```

12-5

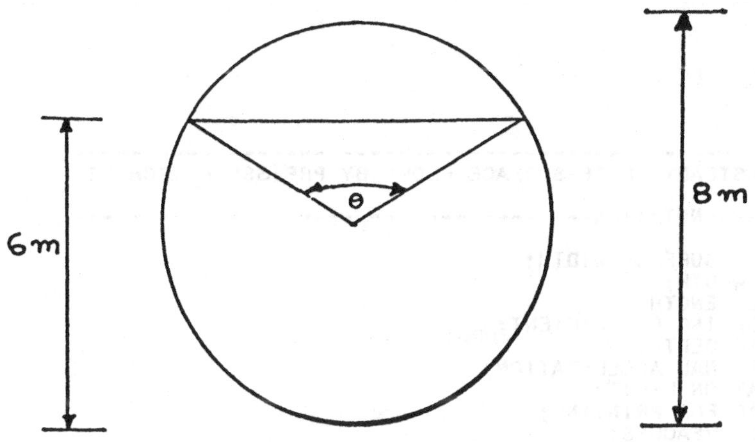

From Eq. 12.27

$$c = \sqrt{gA/B}$$

$$B = 2\sqrt{(4)^2 - (2)^2} = 6.93\ m$$

To calculate area, A , calculate θ first.

$$\theta = 2cos^{-1}\frac{2}{4} = 120^{\circ}$$

$$A = \pi r^2(1 - 120/360) + 0.5 * 2 * 6.93 = 40.44\ m^2$$

$$c = \sqrt{9.81 * 40.44/6.93} = 7.57\ m/s$$

Celerity = 7.57 m/s .

12-6

(i) Gate located at the downstream end : Celerity = c = $\sqrt{gA/B}$

B = 10 m, A = 10*5 = 50 m^2, c = $\sqrt{9.81 * 5}$ = 49.1 m/s .

Absolute velocity of the surge wave = V_w = V-c

or V_w = 4 - 49.1 = - 45.1 m/s. Negative sign indicates that the wave is moving upstream.

(ii) Gate located at the upstream end :

$$V_w = V + c = 4.0 + 49.1 = 53.1\ m/s$$

(iii) Gate located at mid-length of the channel :

One wave moves upstream with an absolute velocity of 45.1 m/s and the other moves downstream with an absolute velocity of 53.1 m/s .

```
C
C-----------------------------------------------------------------------
C   COMPUTATION OF UNSTEADY, FREE-SURFACE FLOWS BY PREISSMANN SCHEME
C
C-----------------------NOTATION-----------------------------------------
C         A = AREA;
C         B = TOP WATER-SURFACE WIDTH;
C         BO = BOTTOM WIDTH;
C         CL = CHANNEL LENGTH;
C         ALPHA = WEIGHTING COEFFICIENT;
C         D = HYDRAULIC DEPTH;
C         G = GRAVITATIONAL ACCELERATION;
C         ITOT = ITERATION LIMIT;
C         KPRINT = FLAG FOR PRINTING;
C         N = NUMBER OF REACHES;
C         P = WETTED PERIMETER;
C         R = HYDRAULIC RADIUS;
C         RN = ROUGHNESS COEFFICIENT;
C         SO = CHANNEL-BOTTOM SLOPE;
C         SS = CHANNEL-SIDE SLOPE;
C         TCLOSE = VALVE CLOSURE TIME;
C         TLAST = TIME UPTO WHICH COMPUTATIONS ARE CARRIED OUT;
C         VFINAL = FINAL VELOCITY AT DOWNSTREAM END;
C         V = VELOCITY;
C         Y = DEPTH OF FLOW;
C         YINI = INITIAL UNIFORM DEPTH;
C
C         VARIABLES JLM, JLN, IA, IGDT, WKAREA, IER ARE RELATED TO
C         THE SUBROUTINE 'LEQT1F' USED FOR SOLVING THE MATRIX.
C
C-----------------------------------------------------------------------
          DIMENSION Y(90),V(90),YNN(90),VNN(90),YN(90),VN(90),D(90)
          DIMENSION DNN(90),SF(90),SFNN(90),CCM(85,85),BBM(85)
          REAL CM(80,80),BM(80),WKAREA(6400)
          INTEGER JLM,JLN,IA,IGDT,IER
          READ(5,*) YINI,SO,RN,CL,ALPHA
          READ(5,*) VFINAL,BO,SS,TCLOSE
          READ(5,*) N,DT,TLAST,KPRINT,G,ITOT
C-----------------------------------------------------------------------
C              INITIAL STEADY-STATE CONDITIONS
C-----------------------------------------------------------------------
          N1=N+1
          DX=CL/N
          DO 10 I=1,N1
          Y(I)=YINI
          A=(BO+SS*Y(I))*Y(I)
          WP=BO+2.0*YINI*SQRT(1+SS*SS)
          HR=(A/WP)**0.6667
          V(I)=(HR/RN)*SQRT(SO)
10        CONTINUE
          VINI=V(N1)
C-----------------------------------------------------------------------
```

```
C             UNSTEADY-STATE COMPUTATIONS
C-----------------------------------------------------------------------
      WRITE(6,1021)
1021  FORMAT(6X,'TIME',6X,'Y(1)',5X,'Y(11)',5X,'Y(21)',5X,'Y(31)'
     * ,5X,'Y(41)')
      WRITE(6,1022)
1022 FORMAT('-----------------------------------------------------------
     *----------')
      T=0.0
20    KFLAG=0
      WRITE(6,1010) T,Y(1),Y(11),Y(21),Y(31),Y(41)
1010  FORMAT(6F10.3)
30    T=T+DT
      IF(T.GT.TLAST) GO TO 999
      KFLAG=KFLAG+1
      IF(TCLOSE.EQ.0.0) GO TO 31
      VLAST=VINI-(VINI-VFINAL)*T/TCLOSE
31    IF(T.GE.TCLOSE) VLAST=VFINAL
C-----------------------------------------------------------------------
C             CALCULATE D(I) AND SF(I)
C-----------------------------------------------------------------------
      DO 40 I=1,N1
      A=(BO+SS*Y(I))*Y(I)
      B=BO+2.0*SS*Y(I)
      P=BO+2.0*Y(I)*SQRT(SS*SS+1.0)
      D(I)=A/B
      R=(A/P)**1.33333
      SF(I)=V(I)*V(I)*RN*RN/R
40    CONTINUE
C-----------------------------------------------------------------------
C             ESTIMATE Y(I) AND V(I) AT THE NEW TIME LEVEL
C-----------------------------------------------------------------------
      DO 50 I=1,N1
      VNN(I)=V(I)
      YNN(I)=Y(I)
      DNN(I)=D(I)
      SFNN(I)=SF(I)
50    CONTINUE
C-----------------------------------------------------------------------
      IFLAG=1
55    CONTINUE
C-----------------------------------------------------------------------
C           ITERATION FOR IMPROVED VALUES
C-----------------------------------------------------------------------
      YNN(1)=YINI
      YN(1)=YINI
      VNN(N1)=VLAST
      VN(N1)=VLAST
      SFNN(N1)=0.0
      JJ=2*N1
      DO 600 KK=1,JJ
      DO 600 LL=1,JJ
      CM(KK,LL)=0.0
      BM(KK)=0.0
600   CONTINUE


      DO 60 I=1,N
      C1=(Y(I)+Y(I+1))/(2.0*DT)
      C2=((1.0-ALPHA)/DX)*(V(I+1)-V(I))
      C3=((1.0-ALPHA)/2.0)*(D(I+1)+D(I))
      C4=((1.0-ALPHA)/DX)*(Y(I+1)-Y(I))
      C5=((1.0-ALPHA)/2.0)*(V(I+1)+V(I))
      C6=(V(I)+V(I+1))/(2.0*DT)
      C7=((1.0-ALPHA)/2.0)*(SF(I+1)+SF(I))
      F1=((ALPHA/2.0)*(DNN(I+1)+DNN(I))+C3)*(2.0*ALPHA*DT/DX)
      F2=((ALPHA/2.0)*(VNN(I+1)+VNN(I))+C5)*(2.0*ALPHA*DT/DX)
      F3=C1*2.0*DT-C2*2.0*DT*((ALPHA/2.0)*(DNN(I+1)+DNN(I))+C3)
      F3=F3-C4*2.0*DT*((ALPHA/2.0)*(VNN(I+1)+VNN(I))+C5)
      F4=-2.0*G*DT*C4+2.0*DT*G*SO+2.0*DT*C6
      F4=F4-2.0*DT*C2*((ALPHA/2.0)*(VNN(I+1)+VNN(I))+C5)
      F4=F4-2.0*G*DT*((ALPHA/2.0)*(SFNN(I+1)+SFNN(I))+C7)
      KKK=2*I-1
      CCM(KKK,KKK)=1.0-F2
      CCM(KKK,KKK+1)=-F1
      CCM(KKK,KKK+2)=1.0+F2
      CCM(KKK,KKK+3)=F1
      BBM(KKK)=F3
```

```
      CCM(KKK+1,KKK)=-2.0*ALPHA*G*DT/DX
      CCM(KKK+1,KKK+1)=1.0-F2
      CCM(KKK+1,KKK+2)=2.0*ALPHA*G*DT/DX
      CCM(KKK+1,KKK+3)=1.0+F2
      BBM(KKK+1)=F4
60    CONTINUE
      BM(1)=BBM(1)-YINI*CCM(1,1)
      BM(2)=BBM(2)+YINI*2.0*ALPHA*G*DT/DX
      JJ1=2*N
      BM(JJ1)=BBM(JJ1)-VLAST*CCM(JJ1,JJ1+2)
      BM(JJ1-1)=BBM(JJ1-1)-VLAST*CCM(JJ1-1,JJ1+2)
      DO 700 I1=1,JJ1
      DO 700 K1=1,JJ1
      CM(I1,K1)=CCM(I1,K1+1)
700   CONTINUE
      JKK=JJ1-2
      DO 699 I=3,JKK
      BM(I)=BBM(I)
699   CONTINUE
C-----------------------------------------------------------------
C            SOLVE THE MATRIX
C-----------------------------------------------------------------
      JLN=JJ1
      IA=80
      JLM=1
      IGDT=3
      CALL LEQT1F (CM,JLM,JLN,IA,BM,IGDT,WKAREA,IER)
C-----------------------------------------------------------------
C     LEQT1F IS A ' SAS ' SUBROUTINE FOR SOLVING THE MATRIX
C-----------------------------------------------------------------
      DO 980 I=1,N
      JKV=2*I-1
      JKY=2*I
      VN(I)=BM(JKV)



      YN(I+1)=BM(JKY)
980   CONTINUE
      YN(1)=YINI
      VN(N1)=VLAST
      DIFF1=0.0
      DIFF2=0.0
      DO 100 I=1,N1
      DIFF1=DIFF1+ABS(VN(I)-VNN(I))
      DIFF2=DIFF2+ABS(YN(I)-YNN(I))
100   CONTINUE
      IF(DIFF1.LE.0.05) GO TO 120
      CONTINUE
      DO 110 I=1,N1
      VNN(I)=VN(I)
      YNN(I)=YN(I)
      A=(BO+SS*YNN(I))*YNN(I)
      B=BO+2.0*SS*YNN(I)
      P=BO+2.0*YNN(I)*SQRT(SS*SS+1.0)
      DNN(I)=A/B
      R=(A/P)**1.33333
      SFNN(I)=VNN(I)*VNN(I)*RN*RN/R
110   CONTINUE
      IFLAG=IFLAG+1
      IF(IFLAG.LE.ITOT) GO TO 55
      WRITE(6,1020) IFLAG
1020  FORMAT(5X,'ITERATION FAIL',5X,I2)
      GO TO 999
120   CONTINUE
C-----------------------------------------------------------------
C            RESETTING THE VALUES FOR NEXT COMPUTATION
C-----------------------------------------------------------------
      DO 150 I=1,N1
      Y(I)=YN(I)
      V(I)=VN(I)
150   CONTINUE
      IF(KFLAG.EQ.KPRINT) GO TO 20
      GO TO 30
999   CONTINUE
      STOP
      END
```

INPUT FILE

```
3.0,0.0001,0.025,5000.0,0.65
0.0,10.0,1.5,0.0
40,20.0,2000.0,1,9.81,25
```

OUTPUT FILE

TIME	Y(1)	Y(11)	Y(21)	Y(31)	Y(41)
0.000	3.000	3.000	3.000	3.000	3.000
20.000	3.000	3.000	3.000	3.000	3.319
40.000	3.000	3.000	3.000	3.000	3.332
60.000	3.000	3.000	3.000	3.000	3.335
80.000	3.000	3.000	3.000	3.000	3.341
100.000	3.000	3.000	3.000	3.000	3.346
120.000	3.000	3.000	3.000	3.000	3.351
140.000	3.000	3.000	3.000	3.000	3.356
160.000	3.000	3.000	3.000	2.998	3.360
180.000	3.000	3.000	3.000	3.004	3.365
200.000	3.000	3.000	3.000	2.999	3.370
220.000	3.000	3.000	3.000	2.987	3.375
240.000	3.000	3.000	3.000	2.999	3.380
260.000	3.000	3.000	3.000	3.050	3.384
280.000	3.000	3.000	3.000	3.119	3.389
300.000	3.000	3.000	3.000	3.183	3.394
320.000	3.000	3.000	3.000	3.227	3.398
340.000	3.000	3.000	3.000	3.255	3.403
360.000	3.000	3.000	3.000	3.272	3.408
380.000	3.000	3.000	3.000	3.282	3.412
400.000	3.000	3.000	3.000	3.289	3.417
420.000	3.000	3.000	3.001	3.295	3.421
440.000	3.000	3.000	3.001	3.300	3.426
460.000	3.000	3.000	2.999	3.304	3.431
480.000	3.000	3.000	2.995	3.309	3.435
500.000	3.000	3.000	2.997	3.314	3.440
520.000	3.000	3.000	3.012	3.318	3.444
540.000	3.000	3.000	3.042	3.323	3.449
560.000	3.000	3.000	3.082	3.327	3.453
580.000	3.000	3.000	3.123	3.332	3.458
600.000	3.000	3.000	3.158	3.336	3.462
620.000	3.000	3.000	3.184	3.341	3.467
640.000	3.000	3.000	3.203	3.345	3.471
660.000	3.000	3.000	3.216	3.349	3.476
680.000	3.000	3.000	3.225	3.354	3.480
700.000	3.000	3.000	3.232	3.358	3.484
720.000	3.000	3.000	3.237	3.363	3.489
740.000	3.000	2.998	3.242	3.367	3.493
760.000	3.000	2.998	3.247	3.371	3.498
780.000	3.000	3.000	3.251	3.376	3.502
800.000	3.000	3.009	3.256	3.380	3.506
820.000	3.000	3.025	3.260	3.384	3.511
840.000	3.000	3.049	3.264	3.389	3.515
860.000	3.000	3.075	3.268	3.393	3.519
880.000	3.000	3.101	3.273	3.397	3.523
900.000	3.000	3.123	3.277	3.401	3.528
920.000	3.000	3.140	3.281	3.406	3.532
940.000	3.000	3.154	3.285	3.410	3.536
960.000	3.000	3.164	3.289	3.414	3.540

Different values of time interval are obtained by using a statement DT = DTNEW*COEF for finding the new time level in the program for problem 12.2. Results are shown for two different cases.

```
N = 30 YD =   3.00 M    MN = 0.025 BO =  10.00 M
 SO =0.0001S =   1.5000 L = 5000.00 M
```

COEF= 1 ——

```
T = 370.852 S
 YMID =   3.00
T = 401.756 S
 YMID =   3.00
T = 432.660 S
 YMID =   3.00
T = 463.565 S
 YMID =   3.00
T = 494.469 S
 YMID =   2.84
T = 525.373 S
 YMID =   2.84
T = 556.278 S
 YMID =   2.80
T = 587.182 S
 YMID =   2.80
T = 618.086 S
 YMID =   2.78
T = 648.990 S
 YMID =   2.78
T = 679.895 S
 YMID =   2.77
T = 710.799 S
 YMID =   2.77
T = 741.703 S
 YMID =   2.75
T = 772.608 S
 YMID =   2.75
T = 803.512 S
 YMID =   2.74
T = 834.416 S
 YMID =   2.74
T = 865.321 S
 YMID =   2.73
T = 896.225 S
 YMID =   2.73
```

COEF= ·98 ----

```
T = 380.676 S
 YMID =   3.00
T = 403.966 S
 YMID =   3.00
T = 426.791 S
 YMID =   2.94
T = 449.159 S
 YMID =   2.95
T = 471.080 S
 YMID =   2.89
T = 492.562 S
 YMID =   2.90
T = 513.615 S
 YMID =   2.85
T = 534.247 S
 YMID =   2.86
T = 554.466 S
 YMID =   2.83
T = 574.281 S
 YMID =   2.84
T = 593.699 S
 YMID =   2.81
T = 612.729 S
 YMID =   2.81
T = 631.378 S
 YMID =   2.79
T = 649.655 S
 YMID =   2.80
T = 667.566 S
 YMID =   2.78
T = 685.118 S
 YMID =   2.78
T = 702.320 S
 YMID =   2.77
T = 719.177 S
 YMID =   2.77
```

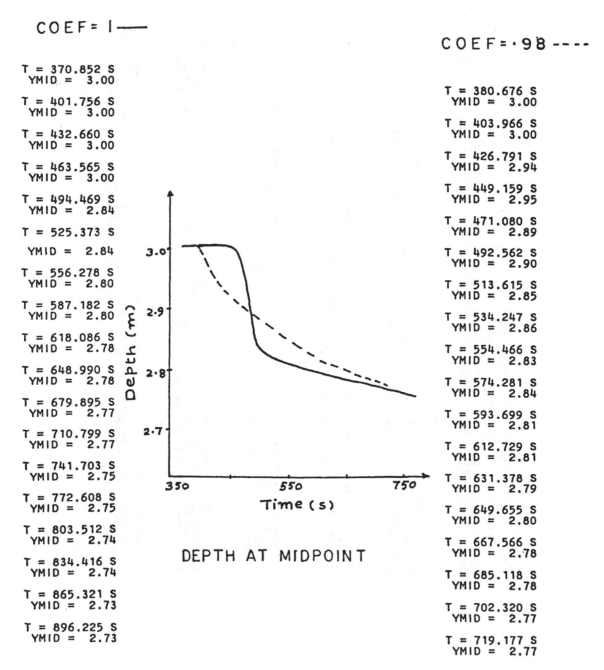

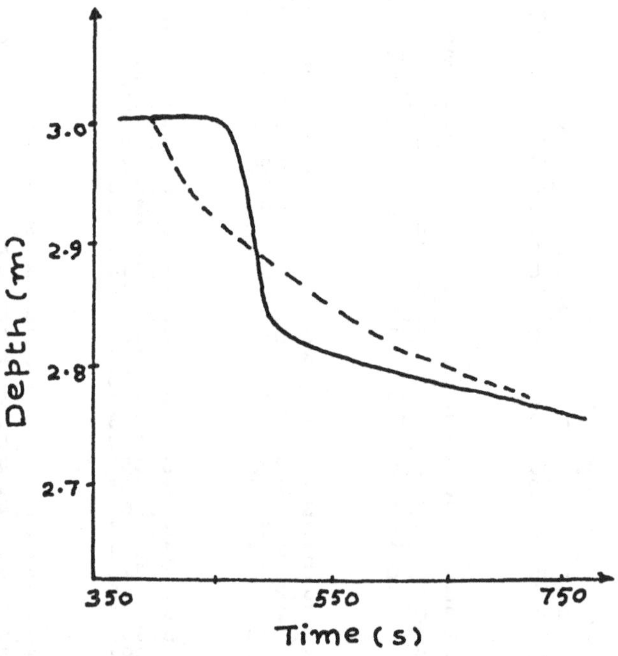

DEPTH AT MIDPOINT

Courant condition is not satisfied if DT = 1.5*DTNEW in the program for problem 12.2 . Results of computation for this case are shown below. As can be seen the computations blow up.

```
N = 10 YD =   3.00 M     MN = 0.025 BO =  10.00 M
 SO =0.0001S =    1.5000 L = 5000.00 M

T =     0.000 S
 Y =   3.00   3.00   3.00   3.00   3.00   3.00   3.00   3.00   3.00   3.00   3.00
  V=   0.65   0.65   0.65   0.65   0.65   0.65   0.65   0.65   0.65   0.65   0.65

T =   92.713 S
 Y =   2.68   3.00   3.00   3.00   3.00   3.00   3.00   3.00   3.00   3.00   3.00
  V=   0.00   0.65   0.65   0.65   0.65   0.65   0.65   0.65   0.65   0.65   0.65

T = 231.783 S
 Y =   2.68   2.63   3.00   3.00   3.00   3.00   3.00   3.00   3.00   3.00   3.00
  V=   0.00  -0.07   0.65   0.65   0.65   0.65   0.65   0.65   0.65   0.65   0.65

T = 370.854 S
 Y =   2.60   2.63   2.58   3.00   3.00   3.00   3.00   3.00   3.00   3.00   3.00
  V=   0.00  -0.07  -0.17   0.65   0.65   0.65   0.65   0.65   0.65   0.65   0.65

T = 509.926 S
 Y =   2.60   2.64   2.58   2.53   3.00   3.00   3.00   3.00   3.00   3.00   3.00
  V=   0.00   0.06  -0.17  -0.30   0.65   0.65   0.65   0.65   0.65   0.65   0.65

T = 648.997 S
 Y =   2.55   2.64   2.68   2.53   2.47   3.00   3.00   3.00   3.00   3.00   3.00
  V=   0.00   0.06   0.15  -0.30  -0.44   0.65   0.65   0.65   0.65   0.65   0.65

T = 788.069 S
 Y =   2.55   2.57   2.68   2.74   2.47   2.41   3.00   3.00   3.00   3.00   3.00
  V=   0.00   0.02   0.15   0.23  -0.44  -0.60   0.65   0.65   0.65   0.65   0.65

T = 927.141 S
 Y =   2.50   2.57   2.59   2.74   2.79   2.41   2.34   3.00   3.00   3.00   3.00
  V=   0.00   0.02   0.03   0.23   0.30  -0.60  -0.79   0.65   0.65   0.65   0.65

T =1066.212 S
 Y =   2.50   2.54   2.59   2.61   2.79   2.85   2.34   2.26   3.00   3.00   3.00
  V=   0.00   0.03   0.03   0.00   0.30   0.33  -0.79  -1.02   0.65   0.65   0.65

T =1205.284 S
 Y =   2.47   2.54   2.58   2.61   2.63   2.85   2.90   2.26   2.17   3.00   3.00
  V=   0.00   0.03   0.05   0.00  -0.05   0.33   0.29  -1.02  -1.31   0.65   0.65

T =1343.008 S
 Y =   2.47   2.51   2.58   2.64   2.63   2.66   2.90   2.92   2.18   2.08   3.00
  V=   0.00   0.01   0.05   0.07  -0.05  -0.14   0.29   0.10  -1.30  -1.67   0.65

T =1473.835 S
 Y =   2.45   2.51   2.55   2.63   2.70   2.67   2.72   2.90   2.87   2.11   3.00
  V=   0.00   0.01   0.00   0.07   0.06  -0.12  -0.23   0.08  -0.38  -1.59  -5.02

T =1550.677 S
 Y =   2.47   2.50   2.56   2.62   2.68   2.76   2.75   2.82   2.71   3.76   3.00
  V=   0.00   0.00   0.02  -0.01   0.02  -0.03  -0.13  -0.36  -0.51  -5.92  -4.31
```

```
T =1618.057 S
 Y =  2.47   2.51   2.56   2.62   2.69   2.74   2.84   2.78   4.42   3.46   3.00
  V=  0.00   0.02  -0.02   0.01  -0.05  -0.04  -0.19  -0.25  -7.02  -4.61  -8.69

T =1673.901 S
 Y =  2.48   2.52   2.57   2.63   2.68   2.79   2.79   4.97   3.79   3.34   3.00
  V=  0.00   0.00   0.01  -0.05  -0.02  -0.15  -0.12  -7.90  -4.52-14.70  -5.99

T =1712.047 S
 Y =  2.50   2.52   2.58   2.63   2.72   2.74   5.04   3.79   4.23   3.38   3.00
  V=  0.00   0.01  -0.03  -0.01  -0.12  -0.07  -7.60  -3.75-23.65  -7.49-30.90

T =1733.093 S
 Y =  2.51   2.54   2.58   2.65   2.69   4.51   3.50   5.48   3.75   3.32   3.00
  V=  0.00  -0.01   0.00  -0.08  -0.05  -5.74  -2.52-30.95  -7.42-63.65  -9.11

T =1744.028 S
 Y =  2.54   2.54   2.60   2.63   3.81   3.17   6.15   3.78   4.40   3.35   3.00
  V=  0.00   0.00  -0.05  -0.03  -3.55  -1.46-30.17  -5.87******  -9.92******

T =1748.737 S
 Y =  2.54   2.57   2.59   3.25   2.92   5.60   3.54   6.15   3.60   3.30   3.00
  V=  0.00  -0.02  -0.01  -1.93  -0.78-21.45  -3.92******  -8.64******-10.16

T =1750.926 S
 Y =  2.58   2.57   2.92   2.76   4.62   3.25   7.14   3.60   4.53   3.30   3.00
  V=  0.00   0.00  -1.00  -0.40-12.85  -2.41******  -6.53******  -9.83******

T =1751.798 S
 Y =  2.57   2.75   2.66   3.81   3.01   6.50   3.43   6.65   3.45   3.31   3.00
  V=  0.00  -0.51  -0.20  -7.12  -1.42-94.12  -4.53******  -8.32******  -9.42

T =1752.196 S
 Y =  2.99   2.61   3.29   2.84   5.33   3.22   7.87   3.45   4.65   3.23   3.00
  V=  0.00  -0.10  -3.84  -0.81-54.20  -2.99******  -6.47******  -8.94******

T =1752.352 S
 Y =  2.66   3.14   2.73   4.34   3.03   7.21   3.34   7.02   3.34   3.31   3.00
  V=  0.00  -1.93  -0.46-29.54  -1.90******  -4.74******  -7.73******  -8.51

T =1752.423 S
 Y =  4.09   2.69   3.75   2.88   5.95   3.18   8.43   3.34   4.72   3.17   3.00
  V=  0.00  -0.23-15.81  -1.18******  -3.33******  -6.24******  -8.13******

T =1752.451 S
 Y =  2.80   3.92   2.79   4.88   3.03   7.81   3.26   7.28   3.25   3.32   3.00
  V=  0.00  -7.92  -0.71******  -2.25******  -4.79******  -7.19******  -7.79

T =1752.464 S
 Y =  8.20   2.79   4.41   2.91   6.52   3.15   8.85   3.26   4.77   3.13   3.00
  V=  0.00  -0.35-74.60  -1.48******  -3.52******  -5.99******  -7.49******

T =1752.469 S
 Y =  2.96   6.30   2.85   5.49   3.03   8.31   3.20   7.46   3.19   3.32   3.00
  V=  0.00-37.34  -0.92******  -2.50******  -4.75******  -6.74******  -7.24
```

```
C
C      COMPUTATION OF UNSTEADY, FREE-SURFACE FLOWS BY LAX'S
C       DIFFUSIVE SCHEME --  MODIFIED FOR PROBLEM 12-8
C
C      ******************** NOTATION **********************
C
C
C      INTERIOR NODES
C
       DO 80 I=2,N
       I1=I-1
       IP1=I+1
       AA=AR(Y(I1))
       PA=WP(Y(I1))
       RA=AA/PA
       SFA=(MN2*V(I1)*V(I1))/(RA**1.333)
       BA=BT(Y(I1))
       AB=AR(Y(IP1))
       BB=BT(Y(IP1))
       P=WP(Y(IP1))
       RB=AB/P
       SFB=(MN2*V(IP1)*V(IP1))/(RB**1.333)
       DM=0.5*(AA/BA + AB/BB)
       SFM=0.5*(SFA+SFB)
       VM=0.5*(V(I1)+V(IP1))
       YM=0.5*(Y(I1)+Y(IP1))
       VP(I)=V(I)-R*G*(Y(IP1)-Y(I1)) - R*VM*(V(IP1)-V(I1))
      1     + G*DT*(SO-SFM)
       YP(I)=Y(I)-R*DM*(V(IP1)-V(I1))-R*VM*(Y(IP1)-Y(I1))
80     CONTINUE
```

OUTPUT

```
N = 10 YD =   3.00 M    MN = 0.025 BO =  10.00 M
 SO =0.0001S =    1.5000 L = 5000.00 M

T =     0.000 S
 Y =   3.00  3.00  3.00  3.00  3.00  3.00  3.00  3.00  3.00  3.00  3.00
  V=   0.65  0.65  0.65  0.65  0.65  0.65  0.65  0.65  0.65  0.65  0.65

T =    92.713 S
 Y =   2.68  3.00  3.00  3.00  3.00  3.00  3.00  3.00  3.00  3.00  3.00
  V=   0.00  0.65  0.65  0.65  0.65  0.65  0.65  0.65  0.65  0.65  0.65

T =   185.426 S
 Y =   2.68  2.86  3.00  3.00  3.00  3.00  3.00  3.00  3.00  3.00  3.00
  V=   0.00  0.39  0.65  0.65  0.65  0.65  0.65  0.65  0.65  0.65  0.65

T =   278.139 S
 Y =   2.65  2.72  2.94  3.00  3.00  3.00  3.00  3.00  3.00  3.00  3.00
  V=   0.00  0.13  0.54  0.65  0.65  0.65  0.65  0.65  0.65  0.65  0.65

T =1855.660 S
 Y =  13.99  4.89  2.79  0.53  1.43  1.61  3.38  4.28  3.49  3.34  3.00
  V=   0.00-28.31-94.57-82.17-95.57-49.02-10.09 -2.28 -7.76-10.89-10.36

T =1857.197 S
 Y =  22.66  4.84  2.58  0.34  1.49  1.53  3.32  4.27  3.52  3.34  3.00
  V=   0.00-52.09******************-78.15-16.55 -2.50 -8.03-11.20-10.60
```